MW00358496

# 3-D Atlas of Stars and Galaxies

Springer
*London*
*Berlin*
*Heidelberg*
*New York*
*Barcelona*
*Hong Kong*
*Milan*
*Paris*
*Santa Clara*
*Singapore*
*Tokyo*

Richard Monkhouse and John Cox

# 3-D ATLAS OF STARS AND GALAXIES

Springer

Richard Monkhouse, MA

John Cox, MA

*Frontispiece – The Hipparcos Satellite Observatory: the 'Star Machine' (Photo ESA)*

ISBN 1-85233-189-5 Springer-Verlag London Berlin Heidelberg

British Library Cataloguing in Publication Data
Monkhouse, Richard, 1950–
3-D atlas of stars and galaxies
1.Stars – Atlases
I.Title II. Cox, John, 1947–
523.8'0223
ISBN 1852331895

Library of Congress Cataloging-in-Publication Data
Monkhouse, Richard.
3-D atlas of stars and galaxies/Richard Monkhouse and John Cox.
p. cm.
Includes bibliographical references.
ISBN 1-85233-189-5 (alk. paper)
1. Stars—Atlases. 2. Galaxies—Atlases. I. Cox, John, 1947- II. Title.
III. Title: Three-D atlas of stars and galaxies.
QB65.M586 1999
523.8'0222'3—dc21 99-26690



Typesetting: The Midlands Book Typesetting Company, Loughborough, UK
Printed and bound by Stamford Press Pte Ltd, Singapore
58/3830-543210 Printed on acid-free paper SPIN 10686997

# Contents

# Introduction

The *3-D Atlas of the Stars and Galaxies* shows the stars and galaxies in three-dimensional space, with the third dimension—distance—directly apparent to stereoscopic vision. The 3-D maps—*stereographs*—are accurate representations constructed from the most recent data available.

Because our eyes are set between two and three inches (50 to 80 mm) apart our stereoscopic vision extends to only a few hundred yards (metres) at most. We use other visual clues, including our experience of how large a particular object ought to appear when observed over a particular range, in order to estimate the distance of things that are further away. We have no direct experience of the size and the distance of stars, and this is why, looking up into the night sky, all the stars appear equally near.

The 3-D maps in the Atlas are unreal to the extent that our eyes are too close together to see parallax over long distances, and the distances involved are, by definition, astronomical. However, they accurately reflect the best measurements and latest estimates of the true distances involved; they show space as it might appear to us if our eyes were *very* much farther apart than they are.

## Distances in Space

Three ways of working out the distance to stars and to galaxies are used in this Atlas. They are trigonometric parallax, spectroscopic parallax, and redshift.

*Trigonometric parallax* uses the orbit of the Earth round the Sun to give widely separated points of view from which the positions of nearby stars can be measured and compared. When measurements are taken a few months apart, the nearer stars show small changes in apparent position. These changes, called 'parallax', are produced by the change in the observer's position as the Earth moves round the Sun. To standardize these measures of parallax, they are adjusted to what they would be if they had been made square-on to the star from either end of a standard baseline of one *astronomical unit*, roughly equivalent to the mean average distance of the Earth from the Sun, 149,600,000 km.

A star about thirty million million kilometres distant (actually 30,857,000,000,000 km; 3·2616 light-years) would show a parallax of one second of arc when observed in this way. One second of arc is 1/3,600 of a degree. Astronomers find it convenient to use a parallax of one second of arc as an expression of distance; this unit is called a *parsec*, abbreviated 'pc'.

A star at twice that distance (and showing a parallax of one-half of a second of arc) is said to be at a distance of 'two parsecs'; one showing a parallax of one-tenth of a second of arc is said to be at a distance of ten parsecs, and so on (1 pc $\approx$3·26 ly; 10 pc $\approx$32·6 ly).

The accuracy of measurements of parallax depends in part on the apparent brightness of the star to be measured, but ground-based photographic detection is routinely capable of resolutions to plus or minus eight *milliarcseconds* (8/3,600,000 of a degree). The satellite observatory *Hipparcos* made measurements to a precision of plus or minus 0·97 milliarcseconds. A star that showed one milliarcsecond of parallax would lie at a distance of 1 kpc (1,000 parsecs, about 3,262 light-years), and this sets a notional limit for distance measurements by trigonometric parallax. The estimated diameter of the Milky Way Galaxy is about 40 kpc, and other methods of determining distance have to be used on galactic and intergalactic scales.

*Spectroscopic parallax* relies on the way that the details of the colour spectrum cast by a star—its 'spectral type'—is a guide to the intrinsic brightness of the star. The distance can be worked out by comparing how bright the star seems to be, with how bright it really is (or is believed to be).

Interpretation of *Redshift* is a third method of estimating distance. The 'redshift of a galaxy' refers to the displacement of absorption lines into the longer wavelengths at the red end of its overall light spectrum. Galaxy redshifts are assumed to be a Doppler effect produced by the speed at which distant galaxies are moving away from the observer, their 'recession velocity'. If the universe is expanding uniformly, the greater the recession velocity, the more distant a galaxy will be. The most redshifted objects shown on the Galaxy Maps are receding at one-tenth of the speed of light, suggesting a distance in the range 500 Mpc.

## The Maps

There are three map series in this book, showing space in three ranges: local space, regional space, and distant space. Coordinates are shown for epoch 2000. The *Near Star Maps* concentrate on stars in the solar neighbourhood, and include a large number of active red dwarfs and inactive white dwarfs, objects too faint to be seen by the naked eye. Most of the objects shown in this map series are nearer than 25 parsecs, eighty light-years.

The *Bright Star Maps* concentrate on the naked-eye stars, often called 'bright stars', that are seen in our region of the Milky Way Galaxy. The Sun is located about 10 kpc (about half-way) out from the centre of the Milky Way system. An extreme supergiant star of an intrinsic luminosity one million times that of the Sun might be just visible to the naked eye over the same distance (10 kpc), and this sets the outside limits of the Bright Star Maps as a 20 kpc circle half the Milky Way in diameter. Most of the objects that are shown on the Bright Star Maps are very much nearer.

The Milky Way Galaxy contains in the region of one hundred thousand million stars, and is one of a similar number of galaxies that make up the visible universe. The *Galaxy Maps* show the galaxies in our region of that universe out to a notional radius of one-tenth of a closed universe model, and concentrate on the structural groupings and strings of galaxies seen in this region.

## The Stereographs

The stereographs of the Near Star Maps show stars as they would appear to someone with eyes about one light-year apart (5,879,000,000,000 miles; 9,460,700,000,000 km). The Bright Star Maps show stars as they might appear to someone with eyes two-and-a-half light-years apart. The Galaxy Maps show the view for someone with eyes five million light-years apart! These figures are based on the notional distance to objects shown at the plane of the page. However, the representation of distances beyond the plane of the page is progressively squeezed up. For example, objects shown on the Galaxy Maps with an assumed distance of 500 Mpc are shown a few inches beyond the plane of the stereograph page. At this range, and at the same but linear scale, a closed universe in proportion to the reader would be reduced to the size of a four-bedroom house.

If you position yourself sixteen inches (400 mm) away from the page, the effective 'depth' that you can see in a comfortable way starts at the plane of the page and extends for six to eight inches 'below' it, producing a little theatre or pocket of virtual space. To accommodate the images into that virtual space, and to get them evenly spaced out, the stereographs use a form of compression similar to a logarithmic scale. To see why this is necessary, consider the case of bright stars.

Bright stars are those visible to the naked eye. They are observed over four magnitudes of distance (distance decades): 1 to 10, 10 to 100, 100 to 1,000, 1,000 to 10,000 parsecs. There are a few stars in the first decades and a few stars in the fourth decade, but most bright stars are seen in the second decade and the near end of the third, in the range ten parsecs out to two hundred. If the distribution were shown in a linear way, most of the stars would be jammed into the first fraction of distance, bunched up into a wall of images a quarter-inch thick. To maximize the differences of distance across the maximum number of objects, the overall distribution has been smoothed out through the available space. The near end of the scale is enlarged, the effective page distance is set a little way into the range, and the distances towards the far end of the scale are progressively compressed.

A different distance and compression scale is used in each map series. The distance of any object can be found by measuring the separation between the red and green images and comparing it to the separations shown on the main key at the beginning of each map series.

## Viewing the 3-D Images

The stereographs should be viewed through red (left eye) and green (right eye) filters, and should be viewed square-on: an imaginary line drawn horizontally across your eyes should be parallel with the width of the page. The best lighting is bright diffuse daylight, so that the page is evenly illuminated and without areas of bright reflection. The comfortable distance for viewing red–green stereographs differs quite a lot from one person to the next. Long-sighted readers may prefer greater viewing distances to those who are short-sighted.

Viewing stereographic images is something that the eyes learn how to do, and the best way is to relax the vision and let the eyes sort it out for themselves. If you have not done it before it might take two or three minutes for the first stereograph to resolve itself. It can help to view the image from a long way away, for example, six feet (2 m). The *Near Star Maps* are the easiest to resolve because the nearer stars are shown larger.

The overall range of depth that is used in the stereographs is set towards the maximum that an experienced viewer will find comfortable, and it is not intended (nor is it possible) that the whole field should be brought into convergence at the same time. This is no different from our ordinary way of seeing things. In ordinary perception we focus our attention on nearby objects or on far-away objects, and the objects we are not concentrating on remain unconverged. The effect is not usually noticed because we are used to it.

It is possible to move down through the depth range by focusing on one object and then on another and nearby object at a slightly greater distance, a sort of 'star-hopping'. The same method can be used to move back up through the field, and this way of viewing has the effect of relaxing the vision and deepening the subjective experience of depth. Once you have seen some part of a first image, the ability to view stereographs improves quickly, and the experience of depth becomes deeper.

# Introduction

# The Near Star Maps

The *Near Star Maps* concentrate on stars within 25 parsecs of the Sun. In interstellar terms, this is our immediate neighbourhood.

A star lying at a distance of 25 parsecs would show 40 milliarcseconds of parallax, easily measured by modern techniques, but even within that 'nearby' range the population of stars remains largely uncatalogued. The difficulty of preparing a catalogue of the nearest stars (and from that catalogue a map) is a matter of faintness and number.

Of 25 stars identified close to the Sun, half are dwarf stars and subdwarf stars of spectral type M: we can suppose that M-type dwarf stars are the most common type of 'nearby' star. A dwarf star of spectral type M5 has an absolute magnitude of 12, which means that it would have an apparent visual magnitude of 12 when observed from 10 parsecs away. There are a million stars in the range of apparent magnitude twelve or brighter: to discover all main-sequence stars down to spectral type M5 within 10 parsecs of the Sun would involve examining the parallax of a million candidate stars.

The near star plots have been made from two catalogues, the *Preliminary Version of the Third Catalogue of Nearby Stars* edited by W. Gliese and H. Jahreiss (1991), and the *Hipparcos Catalogue* (1997).

The *Catalogue of Nearby Stars* is a compilation of ground-based observations intended to list objects within 25 pc of the Sun. The distance information is derived from trigonometric parallax in the case of some stars and from spectroscopic parallax in the case of other stars. It lists 3,000 objects of which the dimmest are main-sequence M8 types and 'degenerate' white dwarfs. To reach completeness the catalogue might need to list more than 8,000 objects, and it would be necessary to examine the parallax of more than a hundred million candidate stars. Some objects are listed on the basis of an incomplete spectral analysis, and the catalogue is likely to include a number of distant giant stars of late spectral types mistaken for nearby dwarf stars. All objects from the catalogue are plotted, but on the annotated maps they are represented by open circles.

The completeness of *Hipparcos* (1997) is defined as (apparent visual magnitude) $V = 7{\cdot}0 - 9{\cdot}0$, with a limiting magnitude of 12·4. Two plots have been taken, and on the annotated maps these plots are distinguished by solid circles. All objects brighter than apparent magnitude 5·25 have been plotted; many are much farther away than 25 pc, but they serve to give a useful 'background' of familiar bright stars. They have been drawn together into constellation figures on the annotated maps. A second plot from Hipparcos shows all objects with a parallax greater than 0·04, that is, all *Hipparcos* objects out to 25 pc. The second plot provides a number of objects not listed in Gliese and Jahreiss, and is a confirmation of a number of objects that are listed.

The diameter of the symbol size seen in the plots is inversely proportional to the distance of the object. There are no set conventions for mapping nearby stars and this way of representing distance is, figuratively speaking, direct. It produces an immediate picture of the location of the nearest stars, at the cost of having the plots of a few more distant objects hidden by the plots of nearer objects.

The annotation of the maps concentrates on nearby objects. Bayer designations are used where possible. Flamsteed numbers (for example, *61 Cygni*) are preferred to other catalogue numberings. The prefixes of catalogue numbers drawn from Gliese and Jahreiss (1991) and shown on the maps have been truncated: prefix 'Gl' (standing for 'Gliese') has been shortened to prefix *G*; prefix 'GJ' shortened to *J*; prefix 'NN' shortened to *N*; and prefix 'Wo' shortened to *W*. Prefix *H* shown on map numberings denotes the *Hipparcos Catalogue* number. Spectral types are taken from Gliese and Jahreiss, and in multiple stars follow the order: A, B, C. An asterisk (*****) indicates that no typing is found for a particular component. Among heterogeneous typings, prefix 'sd' stands for 'subdwarf' (metal-deficient and in theory very old stars), and prefix 'D' stands for 'degenerate' (stars in which nuclear fusion reactions are believed to have ceased).

## Some Interesting Stars

The nearest star to the Sun is *Proxima Centauri* (Near Star Map 6), in many ways a typical nearby star. In Gliese and Jahreiss it is listed as a red dwarf of spectral type M5·5Ve (which means: subtype half-way between M5 and M6, luminosity class V—a main-sequence dwarf star—with peculiarity 'e', emission lines in the spectrum). It is also a member of a multiple-star system. The primary star of that system is *Alpha Centauri*, 'Rigil Kent', itself a double star of which the 'A' component is of the same spectral class as the Sun, G2V. The less massive 'B' companion is a K0V star that reaches an elongation of 17 arcseconds. Observed over a distance of 1·35 pc this separation implies that the A and B components are 23 AU (about 3·4 thousand million km) apart, a similar distance to that of the Sun from planet Uranus. *Proxima Centauri* is separated from *Alpha Centauri* by more than 2° of arc, corresponding to a distance of more than 10,000 AU. If *Proxima Centauri* marks the edge of the system, then the edge extends about 5% of the total distance from *Alpha Centauri* back to the Sun.

In orbital models of multiple-star systems the strong gravity fields throw out low-mass components. It seems that planets are unlikely to be found within multiple-star systems. They are more likely to be found orbiting single stars.

Trying to observe a planet close to another star is like trying to see a pea suspended in front of an arc light, so the usual method of planet searching is to look for small and regular changes in a star's position, indicating unseen companions. A recent such candidate is *Upsilon Andromeda* (map 2), distance 13·5 pc, spectral type F8V, a star not unlike the Sun. However, unseen companions inferred on the basis of a regular wobble in star positions are necessarily more massive than Earth, ranging from about the mass of Jupiter up to so-called 'brown stars' bordering on the 0·08 solar masses needed to start the fusion reactions that power a true star.

*Barnard's Star* (map 5), a red dwarf type M5, distance 1·8 pc, visual magnitude 9·5, holds the 'angular speed record' of more than ten arcseconds a year (just passing through!). At present it is the fourth closest star to the Sun, and perturbations in its track suggest an unseen companion or companions of two to three Jupiter masses.

*61 Cygni*, a naked-eye object of combined magnitude 4·6, is the twelfth closest system, distance 3·5 pc. The major elements are listed in Gliese and Jahreiss as K5Ve and K7Ve, but there is at least one invisible companion at about ten Jupiter masses, perhaps a brown star.

The fifth closest system is *Sirius*, distance 2·6 pc. *Sirius A*, spectral type A1V, is a 'dwarf' star, meaning that the star is seen in an early and compact condition (in some remote future *Sirius A* will expand to become a diffuse 'giant'). *Sirius A* is the brightest star in the night sky, but this is because it is observed at a close distance: in other respects it is a typical bright star. Its companion is the famous *Sirius B*, the first white-dwarf star to be seen (proposed by Bessel in 1844, observed by Clark in 1862), spectral type DA2. White dwarfs are 'degenerate' stars that have ceased nuclear fusion, but continue to radiate light from their reserves of conserved heat. White dwarfs have an absolute magnitude in the range 10 to 15, making them difficult objects to observe. The nearby star *Procyon* (map 3) is another system with a white-dwarf component, *Procyon B.*

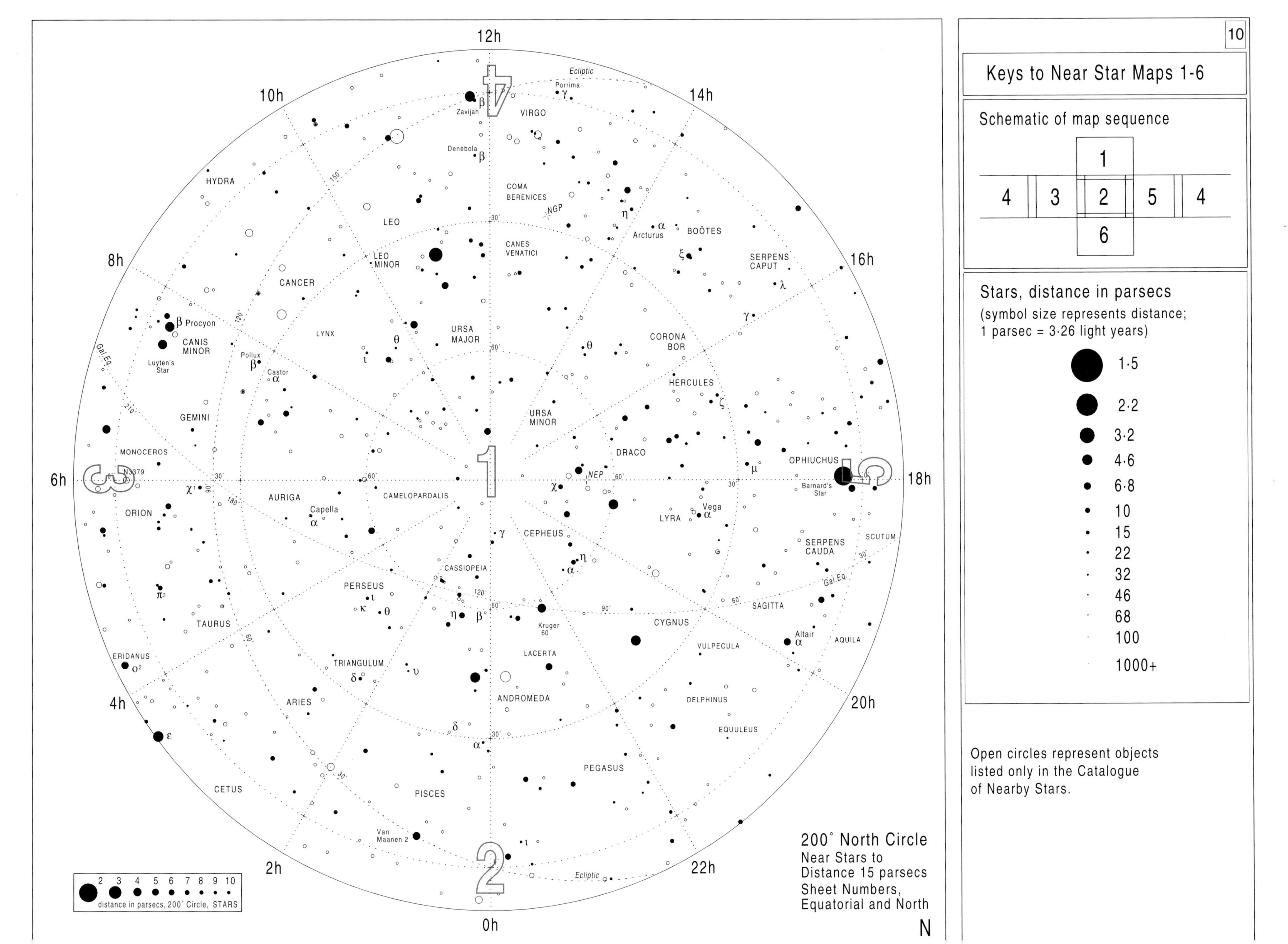

Keys to Near Star Maps 1-6
Schematic of map sequence
1
4 3 2 5 4
6
Stars, distance in parsecs
(symbol size represents distance;
1 parsec = 3·26 light years)
1·5
2·2
3·2
4·6
6·8
10
15
22
32
46
68
100
1000+
Open circles represent objects
listed only in the Catalogue
of Nearby Stars.
200° North Circle
Near Stars to
Distance 15 parsecs
Sheet Numbers,
Equatorial and North
N
12h
14h
16h
18h
20h
22h
0h
2h
4h
6h
8h
10h
Ecliptic
Gal.Eq.
NGP
NEP
VIRGO
COMA BERENICES
BOÖTES
SERPENS CAPUT
CANES VENATICI
LEO
LEO MINOR
HYDRA
CANCER
LYNX
URSA MAJOR
CORONA BOR
HERCULES
URSA MINOR
DRACO
OPHIUCHUS
Barnard's Star
LYRA
Vega
CEPHEUS
SERPENS CAUDA
SCUTUM
SAGITTA
AQUILA
Altair
CYGNUS
VULPECULA
DELPHINUS
EQUULEUS
PEGASUS
LACERTA
Kruger 60
CASSIOPEIA
CAMELOPARDALIS
AURIGA
Capella
PERSEUS
TRIANGULUM
ANDROMEDA
ARIES
PISCES
Van Maanen 2
CETUS
TAURUS
ERIDANUS
ORION
MONOCEROS
N3379
GEMINI
CANIS MINOR
Procyon
Luyten's Star
Pollux
Castor
Porrima
Zavijah
Denebola
Arcturus
distance in parsecs, 200° Circle, STARS
2 3 4 5 6 7 8 9 10

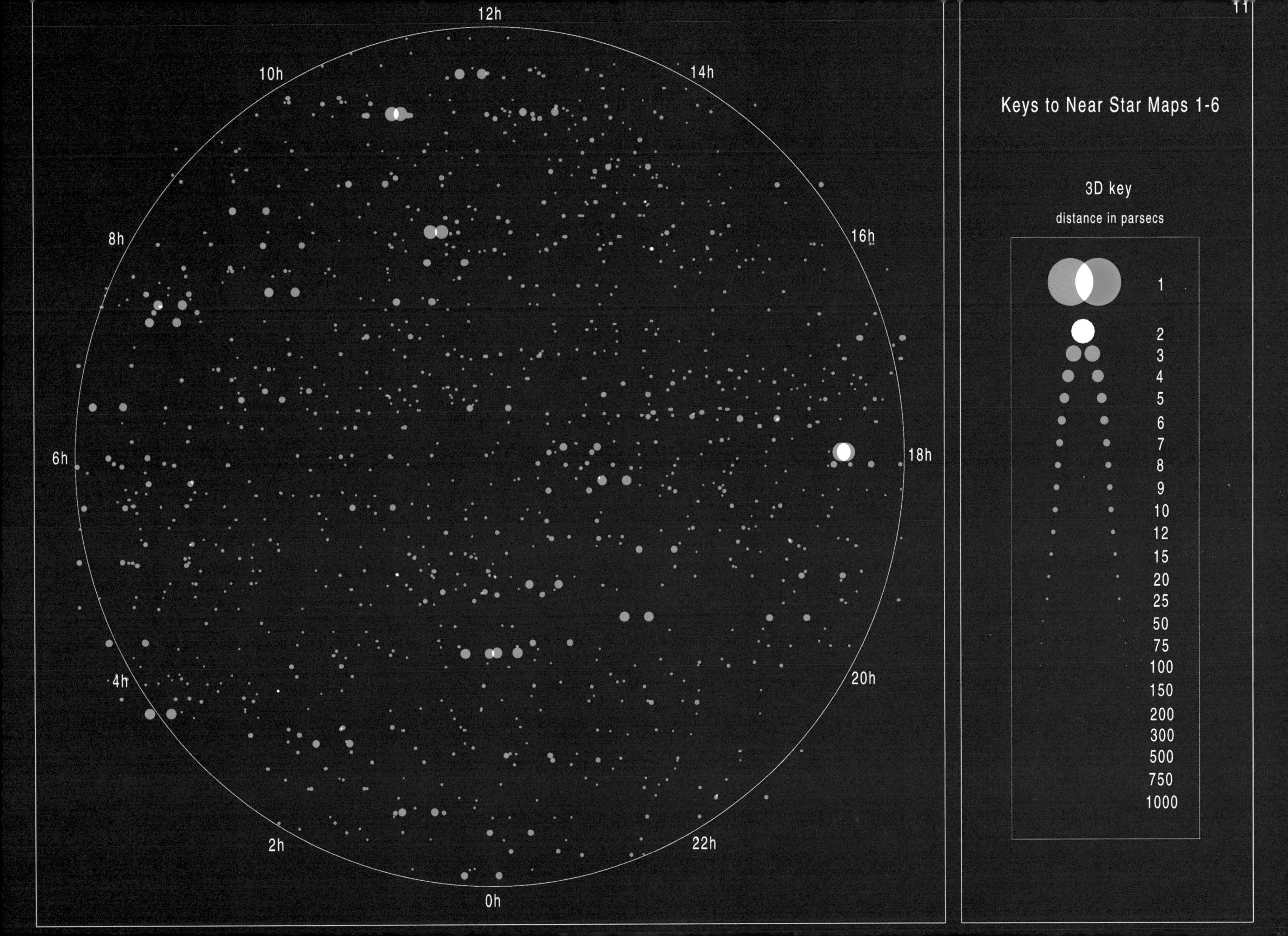

12h
10h
14h
8h
16h
6h
18h
4h
20h
2h
22h
0h
Keys to Near Star Maps 1-6
3D key
distance in parsecs
1
2
3
4
5
6
7
8
9
10
12
15
20
25
50
75
100
150
200
300
500
750
1000

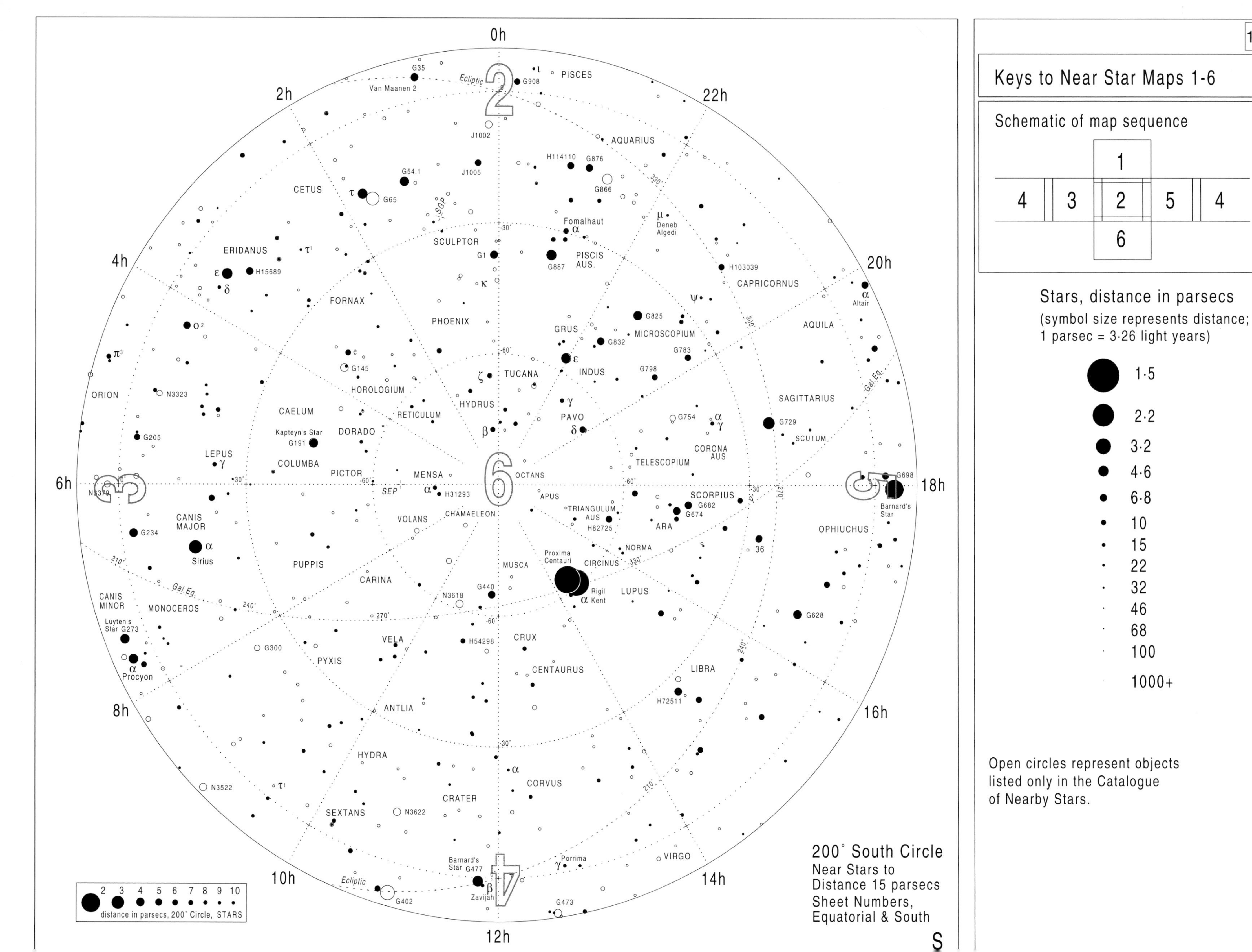
Keys to Near Star Maps 1-6
Schematic of map sequence
1
4 3 2 5 4
6
Stars, distance in parsecs
(symbol size represents distance;
1 parsec = 3·26 light years)
1·5
2·2
3·2
4·6
6·8
10
15
22
32
46
68
100
1000+
Open circles represent objects
listed only in the Catalogue
of Nearby Stars.
200° South Circle
Near Stars to
Distance 15 parsecs
Sheet Numbers,
Equatorial & South
S
0h
2h
4h
6h
8h
10h
12h
14h
16h
18h
20h
22h
PISCES
AQUARIUS
CETUS
ERIDANUS
FORNAX
SCULPTOR
PHOENIX
PISCIS AUS.
CAPRICORNUS
AQUILA
SAGITTARIUS
SCUTUM
GRUS
MICROSCOPIUM
INDUS
TUCANA
HYDRUS
HOROLOGIUM
RETICULUM
DORADO
CAELUM
COLUMBA
PICTOR
MENSA
OCTANS
PAVO
TELESCOPIUM
CORONA AUS
SCORPIUS
OPHIUCHUS
ARA
TRIANGULUM AUS
NORMA
CIRCINUS
LUPUS
APUS
CHAMAELEON
VOLANS
MUSCA
CRUX
CENTAURUS
LIBRA
VIRGO
CORVUS
CRATER
HYDRA
SEXTANS
ANTLIA
PYXIS
VELA
CARINA
PUPPIS
CANIS MAJOR
LEPUS
ORION
CANIS MINOR
MONOCEROS
Van Maanen 2
G35
G908
J1002
J1005
G54.1
G65
H114110
G876
G866
Fomalhaut
Deneb Algedi
G1
G887
H103039
Altair
H15689
G825
G832
G783
G798
G754
G729
G698
Barnard's Star
G145
N3323
G205
Kapteyn's Star G191
H31293
SEP
H82725
G682
G674
36
Proxima Centauri
Rigil Kent
G440
N3618
H54298
G628
H72511
G234
Sirius
Luyten's Star G273
Procyon
G300
N3522
N3622
Barnard's Star G477
Zavijah
G402
G473
Porrima
N3370
Ecliptic
Gal.Eq.
SGP
30°
60°
210°
240°
270°
300°
330°
2 3 4 5 6 7 8 9 10
distance in parsecs, 200° Circle, STARS

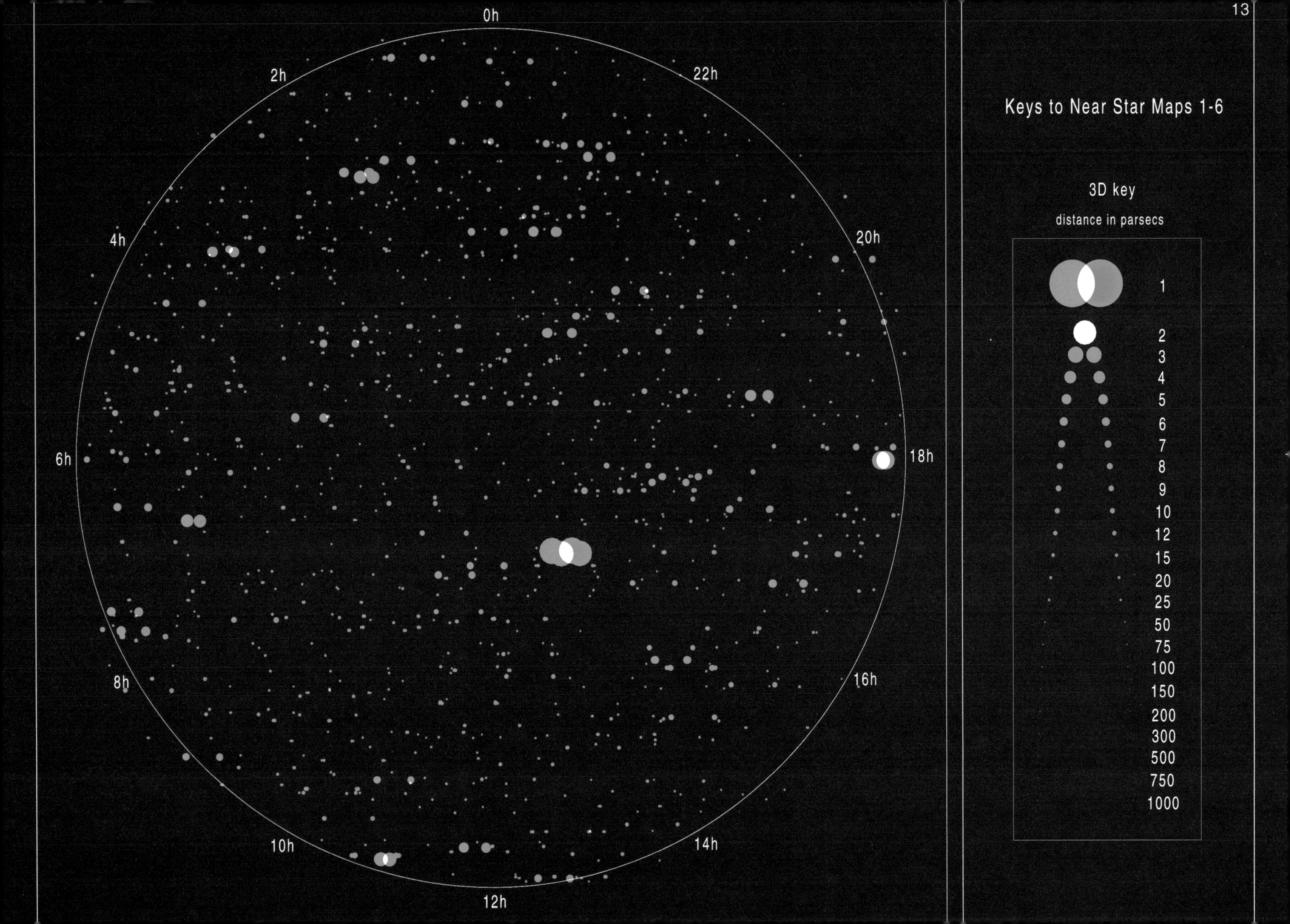
0h
2h
22h
4h
20h
6h
18h
8h
16h
10h
14h
12h
Keys to Near Star Maps 1-6
3D key
distance in parsecs
1
2
3
4
5
6
7
8
9
10
12
15
20
25
50
75
100
150
200
300
500
750
1000

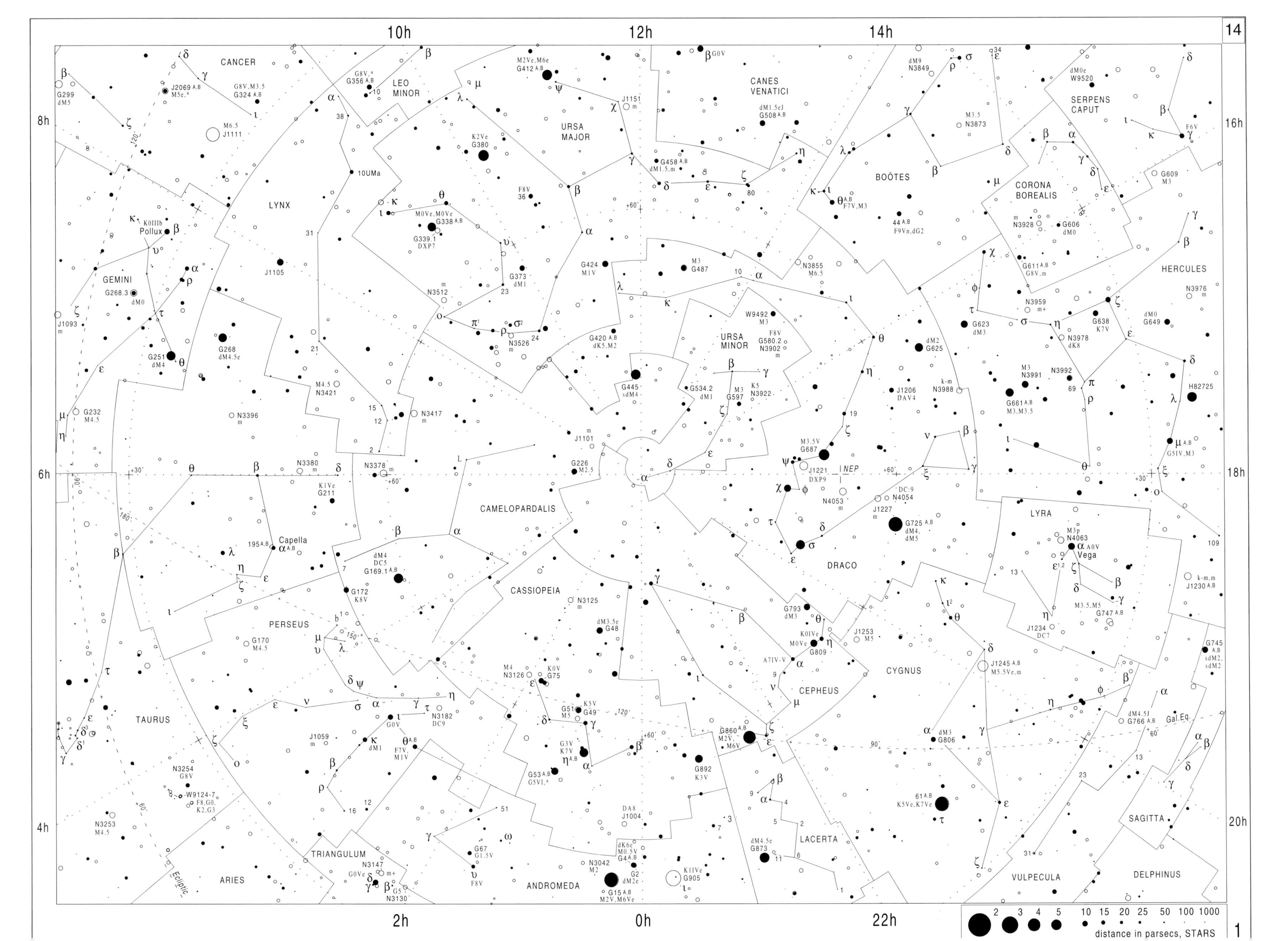
10h
12h
14h
8h
16h
6h
18h
4h
20h
2h
0h
22h
CANCER
LEO MINOR
URSA MAJOR
CANES VENATICI
SERPENS CAPUT
BOÖTES
CORONA BOREALIS
HERCULES
LYNX
GEMINI
URSA MINOR
CAMELOPARDALIS
DRACO
LYRA
Vega
Pollux
Capella
PERSEUS
CASSIOPEIA
CEPHEUS
CYGNUS
TAURUS
TRIANGULUM
ARIES
ANDROMEDA
LACERTA
SAGITTA
VULPECULA
DELPHINUS
NEP
Ecliptic
Gal.Eq.
distance in parsecs, STARS
2 3 4 5 10 15 20 25 50 100 1000

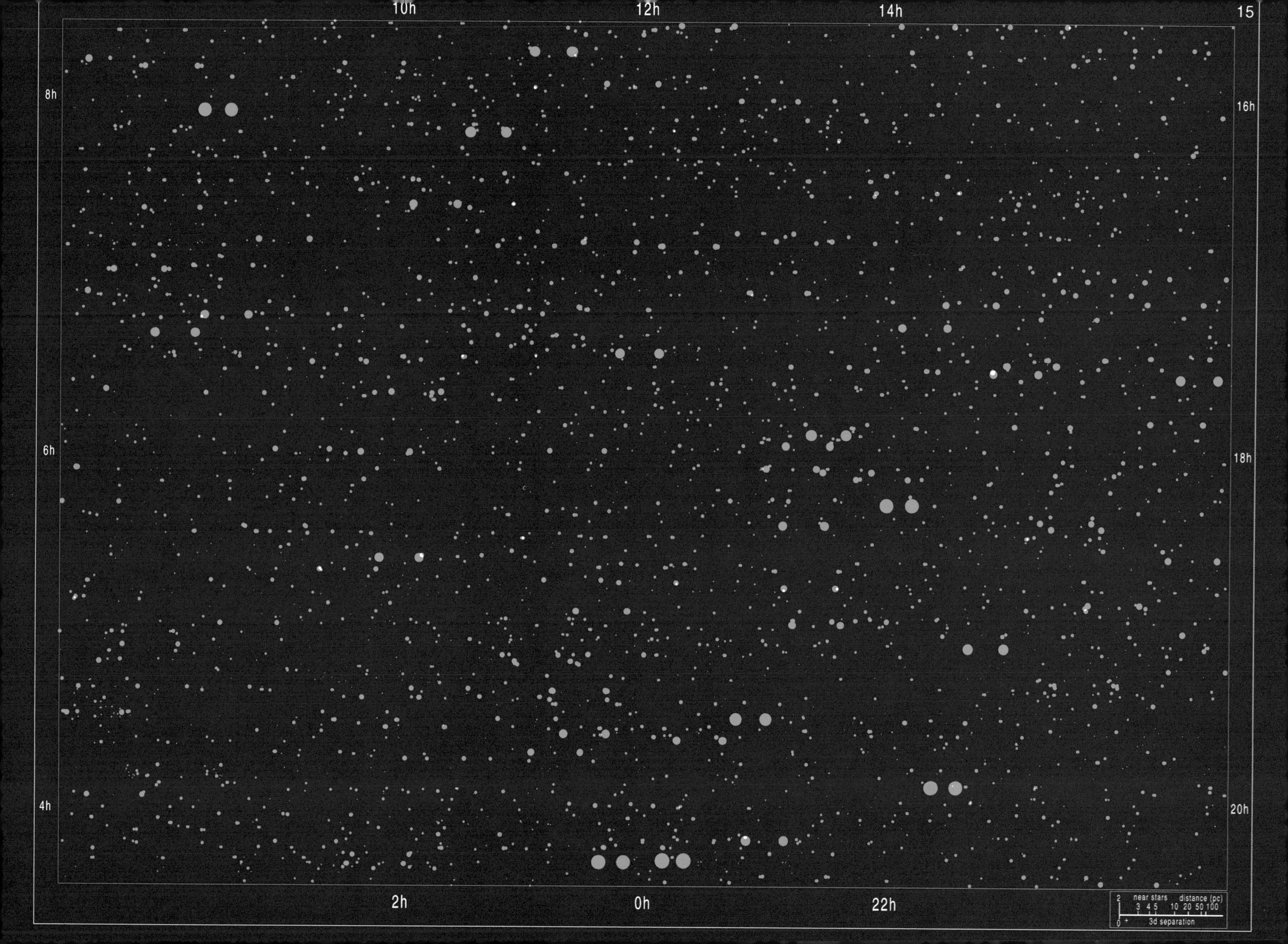
10h
12h
14h
8h
6h
4h
16h
18h
20h
2h
0h
22h
near stars
distance (pc)
2 3 4 5 10 20 50 100
0 1
3d separation

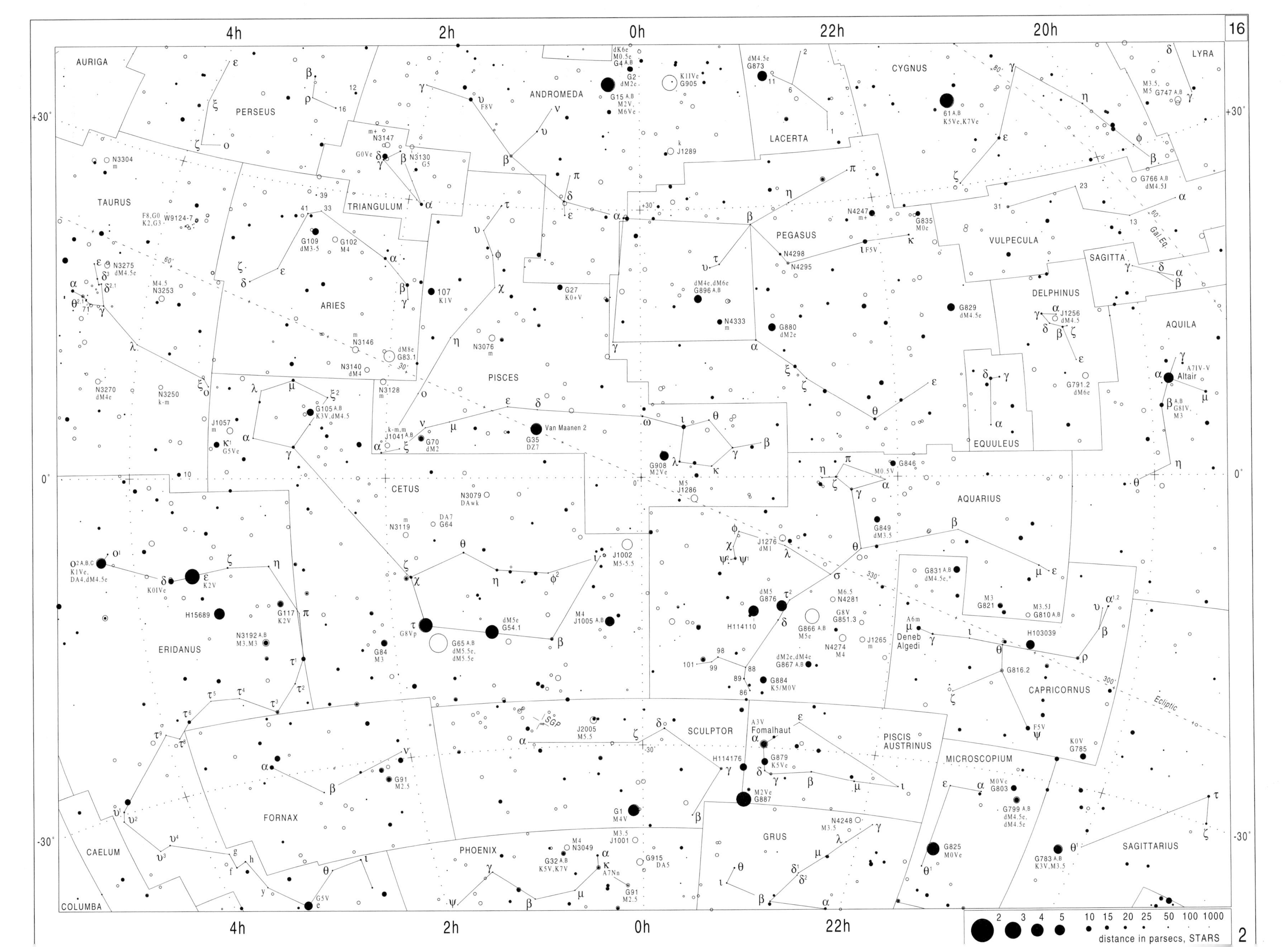
4h
2h
0h
22h
20h
+30°
0°
-30°
AURIGA
PERSEUS
ANDROMEDA
LACERTA
CYGNUS
LYRA
TAURUS
TRIANGULUM
ARIES
PEGASUS
VULPECULA
SAGITTA
DELPHINUS
AQUILA
PISCES
EQUULEUS
CETUS
AQUARIUS
ERIDANUS
CAPRICORNUS
SCULPTOR
PISCIS AUSTRINUS
MICROSCOPIUM
FORNAX
PHOENIX
GRUS
SAGITTARIUS
CAELUM
COLUMBA
Altair
Van Maanen 2
Fomalhaut
Deneb Algedi
Gal.Eq.
Ecliptic
SGP
61 A,B
K5Ve,K7Ve
G35
DZ7
distance in parsecs, STARS
2 3 4 5 10 15 20 25 50 100 1000

4h
2h
0h
22h
20h
+30°
0°
-30°
near stars
distance (pc)
2 3 4 5 10 20 50 100
0
3d separation

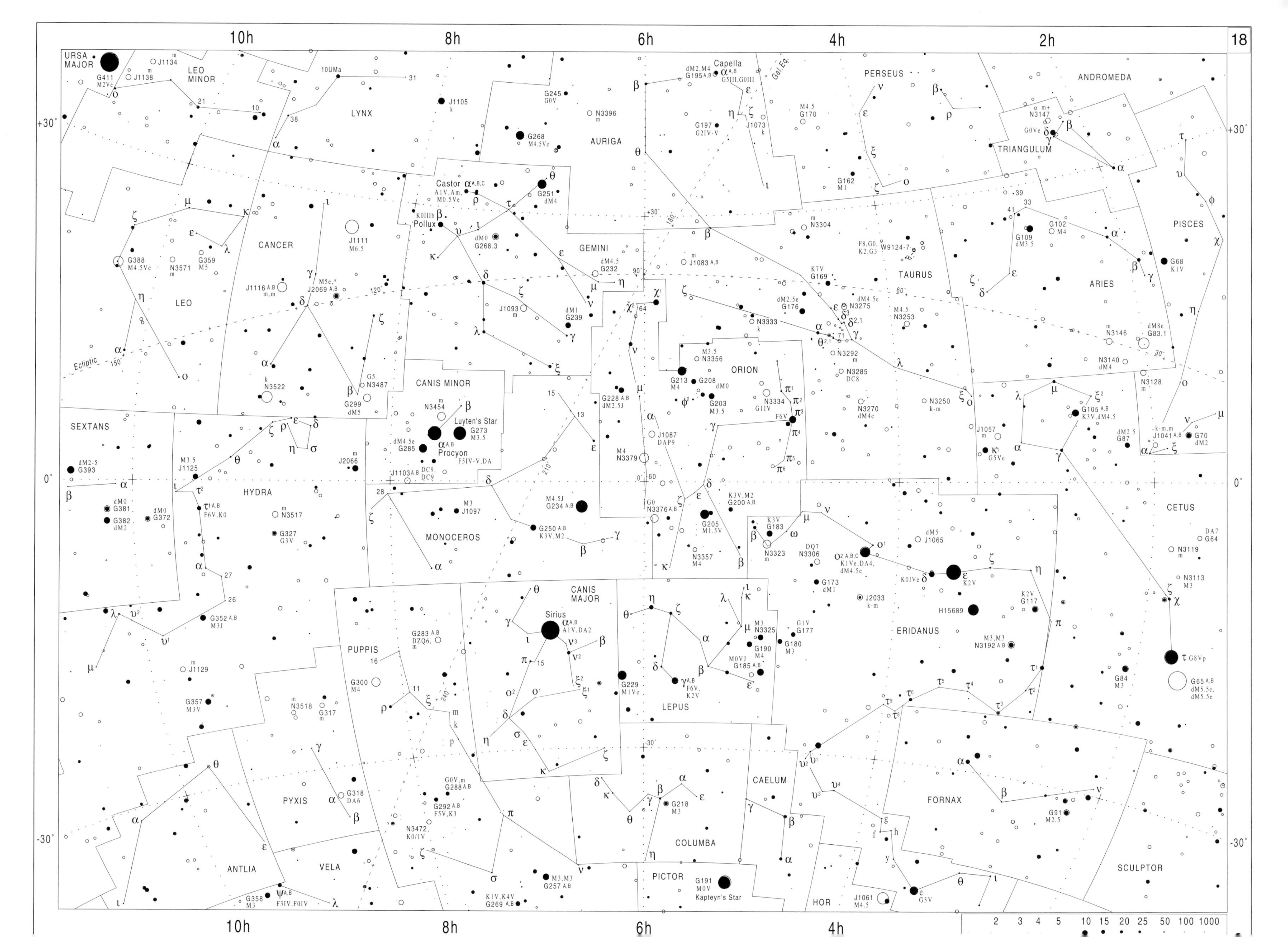

10h
8h
6h
4h
2h
+30°
0°
-30°
URSA MAJOR
LEO MINOR
LYNX
AURIGA
Capella
PERSEUS
ANDROMEDA
TRIANGULUM
PISCES
CANCER
Castor
Pollux
GEMINI
TAURUS
ARIES
LEO
Ecliptic
Gal. Eq.
CANIS MINOR
Procyon
Luyten's Star
ORION
SEXTANS
HYDRA
MONOCEROS
CETUS
CANIS MAJOR
Sirius
LEPUS
ERIDANUS
PUPPIS
PYXIS
ANTLIA
VELA
COLUMBA
CAELUM
PICTOR
Kapteyn's Star
HOR
FORNAX
SCULPTOR
2 3 4 5 10 15 20 25 50 100 1000

10h
8h
6h
4h
2h
+30°
0°
-30°
near stars
distance (pc)
2 3 4 5 10 20 50 100
0
3d separation

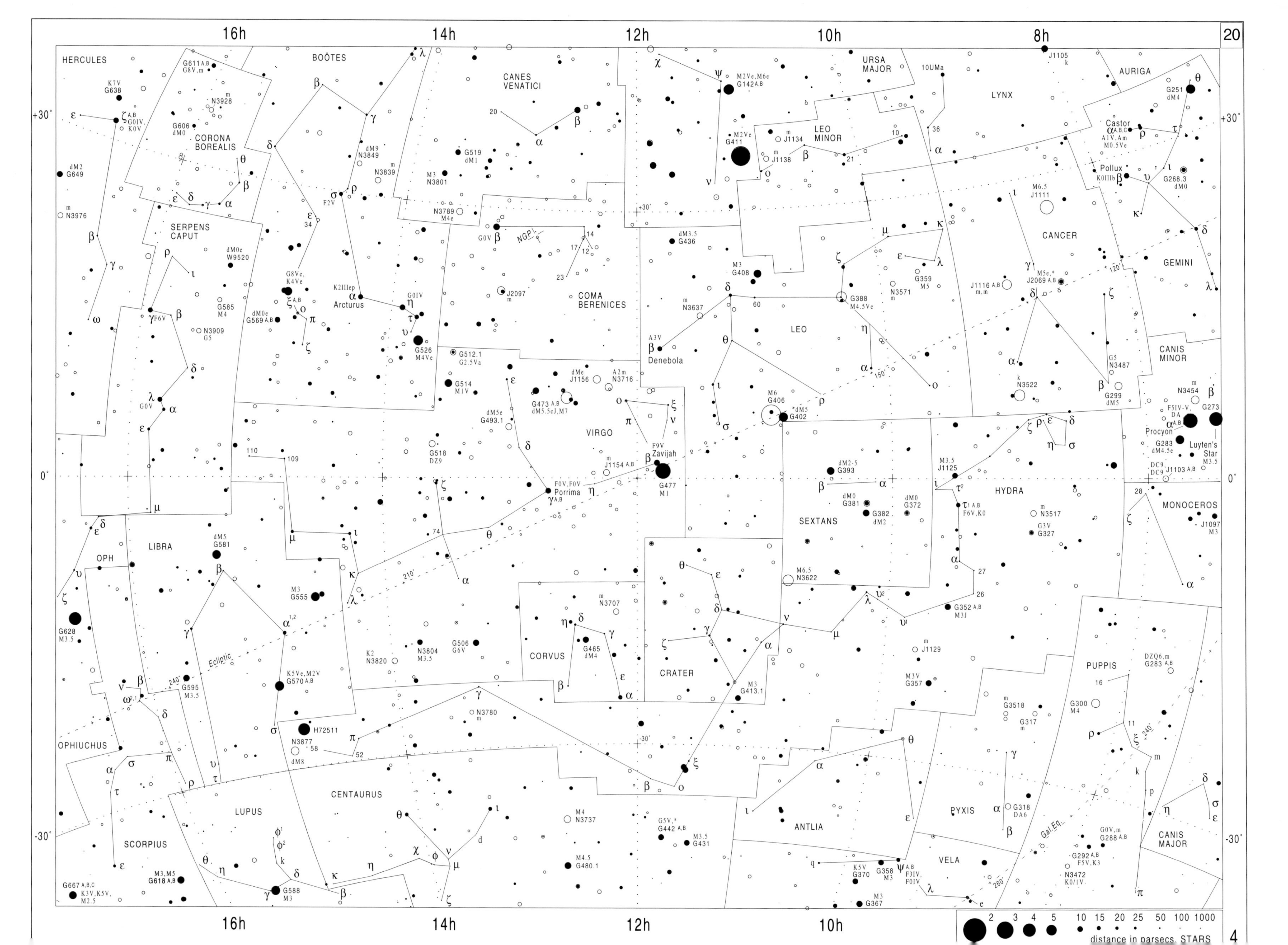

16h
14h
12h
10h
8h
+30°
0°
-30°
HERCULES
BOÖTES
CANES VENATICI
URSA MAJOR
LYNX
AURIGA
LEO MINOR
CORONA BOREALIS
SERPENS CAPUT
COMA BERENICES
CANCER
GEMINI
LEO
CANIS MINOR
VIRGO
HYDRA
SEXTANS
MONOCEROS
LIBRA
OPH
CORVUS
CRATER
PUPPIS
OPHIUCHUS
CENTAURUS
LUPUS
ANTLIA
PYXIS
SCORPIUS
VELA
CANIS MAJOR
Arcturus
Denebola
Castor
Pollux
Procyon
Luyten's Star
Porrima
Zavijah
Ecliptic
Gal.Eq.
NGP
STARS
distance in parsecs
2 3 4 5
10 15 20 25 50 100 1000

16h
14h
12h
10h
8h
+30°
0°
-30°
near stars
distance (pc)
2 3 4 5 10 20 50 100
0
3d separation

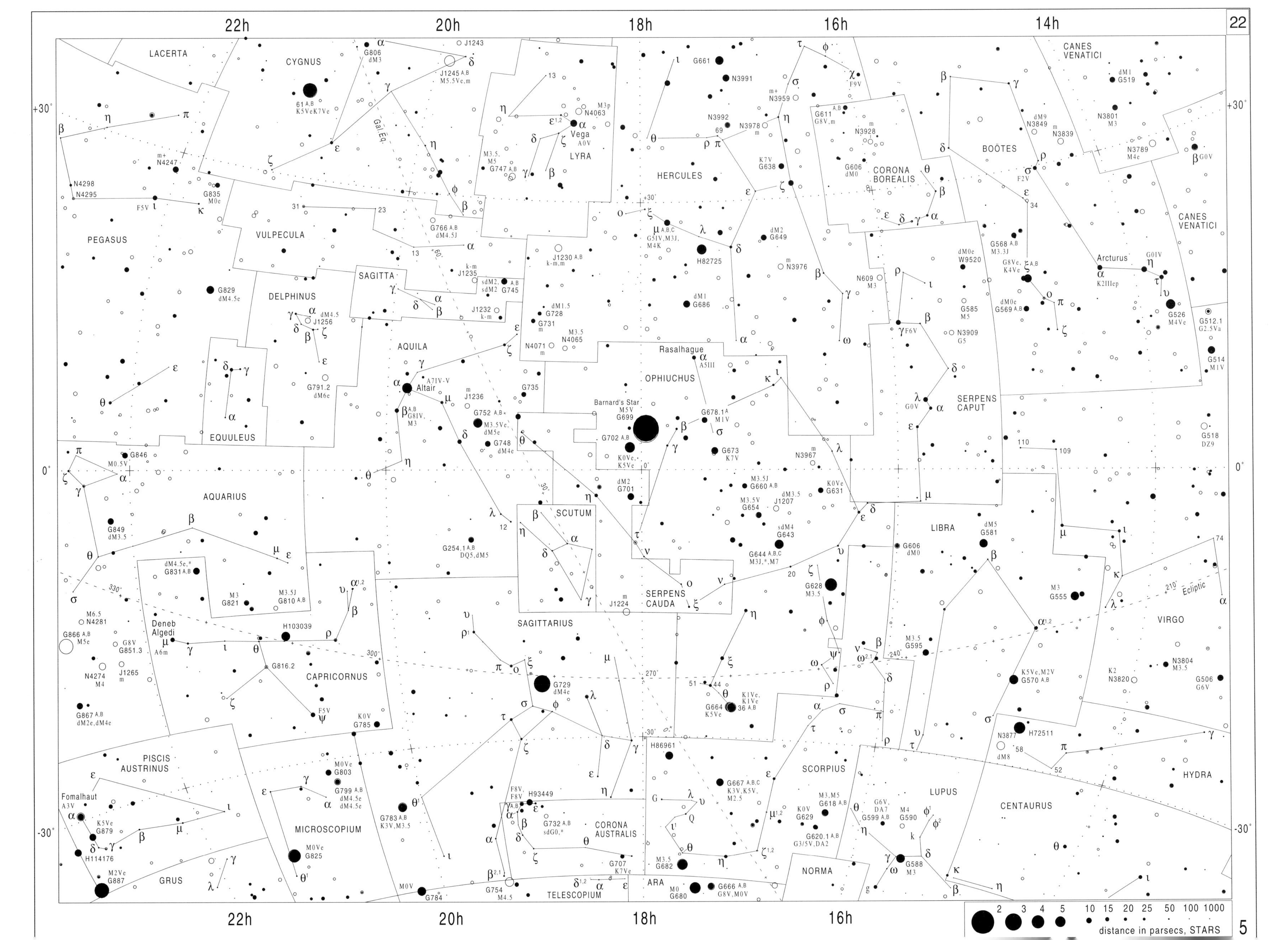

22
5
22h
20h
18h
16h
14h
+30°
0°
-30°
LACERTA
CYGNUS
LYRA
HERCULES
CORONA BOREALIS
BOÖTES
CANES VENATICI
PEGASUS
VULPECULA
SAGITTA
DELPHINUS
AQUILA
EQUULEUS
OPHIUCHUS
SERPENS CAPUT
AQUARIUS
SCUTUM
LIBRA
SERPENS CAUDA
SAGITTARIUS
CAPRICORNUS
VIRGO
PISCIS AUSTRINUS
MICROSCOPIUM
CORONA AUSTRALIS
SCORPIUS
LUPUS
CENTAURUS
HYDRA
GRUS
TELESCOPIUM
ARA
NORMA
Vega
Arcturus
Rasalhague
Barnard's Star
Altair
Deneb Algedi
Fomalhaut
Gal. Eq.
Ecliptic
2 3 4 5 10 15 20 25 50 100 1000
distance in parsecs, STARS

22h
20h
18h
16h
14h
+30°
0°
-30°
near stars
2
3 4 5
distance (pc)
10 20 50 100
0
3d separation

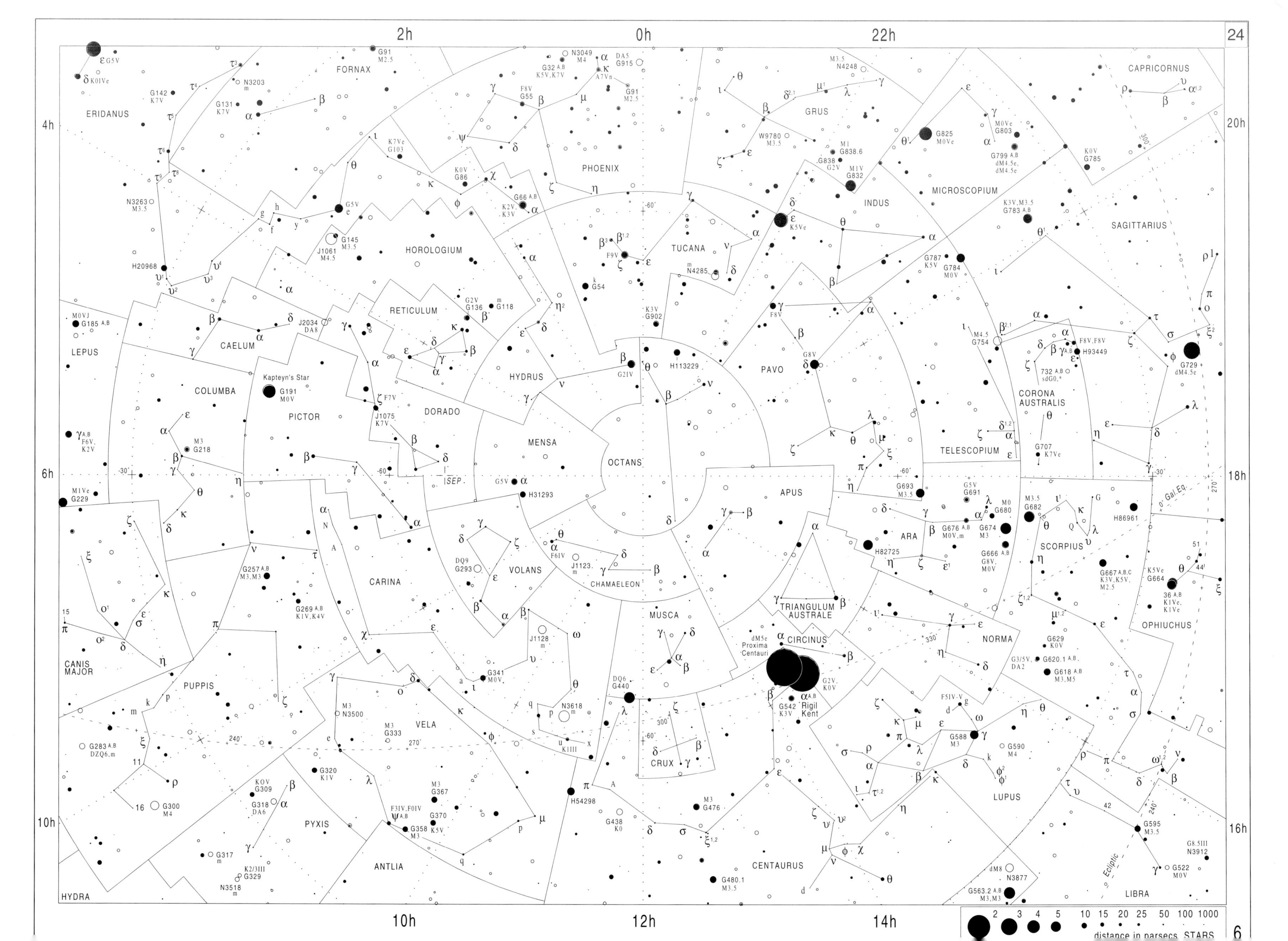
2h
0h
22h
20h
4h
6h
10h
18h
16h
10h
12h
14h
CAPRICORNUS
FORNAX
ERIDANUS
PHOENIX
GRUS
SAGITTARIUS
MICROSCOPIUM
INDUS
HOROLOGIUM
TUCANA
RETICULUM
HYDRUS
PAVO
CORONA AUSTRALIS
TELESCOPIUM
LEPUS
CAELUM
COLUMBA
PICTOR
DORADO
MENSA
OCTANS
APUS
ARA
SCORPIUS
OPHIUCHUS
VOLANS
CHAMAELEON
CARINA
MUSCA
TRIANGULUM AUSTRALE
CIRCINUS
NORMA
CANIS MAJOR
PUPPIS
VELA
CRUX
LUPUS
CENTAURUS
PYXIS
ANTLIA
HYDRA
LIBRA
Kapteyn's Star
Proxima Centauri
Rigil Kent
Gal Eq.
Ecliptic
SEP
distance in parsecs STARS

2h
0h
22h
4h
20h
6h
18h
10h
16h
10h
12h
14h
near stars
2
3 4 5
distance (pc)
10 20 50 100
0
3d separation

# Introduction

# The Bright Star Maps

The Bright Star Maps are plotted from the *Hipparcos Catalogue* (1998) to a limit at visual magnitude 6·49. Non-stellar objects to a limit at magnitude 6·49 have been shown on the annotated maps and are plotted from the catalogue *NGC 2000.0* (1989). Non-stellar objects are not shown on the stereographs.

The European Space Agency satellite *Hipparcos* was launched by Ariane 4 on August 8th 1989 and made between 100 and 150 separate observations of the positions of more than 100,000 stars in the period November 1989 to March 1993. The satellite made observations free of the distorting effects of the Earth's atmosphere, and made them in two directions at the same time to give an independent frame of reference for the measurements. The reported median precision was plus or minus 0·97 milliarcseconds.

*Hipparcos* measurements are taken into the stereographs without intervention. There is a distance scale next to the 'North' and 'South' circles at the beginning of the map series (a colour photocopy of this key would make a useful bookmark). Sample distances given on the annotated maps are rounded out to two figures. They allow another way of reading off the distance shown in the stereographs. The milliarcsecond error that remains in the *Hipparcos* measurements means that the distances shown or quoted to remote stars should be regarded with caution.

The page distance of the stereographs is set at 5 pc. You may notice that there is considerable compression of depth for objects represented at distances greater than 100 pc. In the annotated maps, Bayer designations are preferred where possible, and Flamsteed numbers are preferred to Argelander alphanumerics.

## Some Interesting Objects

A point often made about the depth distribution of stars is that constellation groupings are for the most part imaginary, that the stars of a seeming group like the Big Dipper are at different distances and have no association in three-dimensional space. As it happens, the stars of the Big Dipper do not illustrate this effect (Bright Star Map 1). The five central stars, *Zeta, Epsilon, Delta, Gamma, Beta,* are all A-type main-sequence stars at distances close to 25 pc, a remarkable coincidence of star type and distance.

Undisputed associations in three-dimensional space are observed in open clusters such as the *Hyades* and the *Pleiades* (map 4). The *Hyades* themselves are a lovely example of a moving cluster (see special view on p. 91), with *Aldebaran* a false member, moving across the field half-way between the Sun and the true grouping.

A looser kind of grouping is seen in the depiction of stars in the Milky Way. The plane of the Milky Way galaxy is represented by the *Galactic Equator*, with the centre of the galaxy in direction 0° of the galactic equator beyond the stars of *Sagittarius* (maps 13 and14), and with the edge of the galaxy lying in direction 180° beyond the stars of *Auriga* (maps 4 and 5). It is established that the galaxy has the shape of a flattened spiral with a number of curving arms defined by dust clouds, star-formation, and dense populations of highly luminous young stars. Star clouds produce a noticeable thickening along the line of the galactic plane that can be seen crossing some part of most of the maps. They represent the near part of the major arms of the galaxy region, with the 'Sagittarius Arm' lying about 2 kpc towards the centre of the galaxy, and the 'Perseus Arm' lying about 2 kpc towards the edge.

The Sun is located south of the galactic plane, and a trailing arm-like element, the 'Orion Arm', is seen separated from the galactic plane in the region of *Perseus* and along a line through the *Pleiades,* the *Hyades,* and *Orion* (map 4). Particularly clear areas of galactic structure are seen in the galactic plane through *Centaurus* and *Vela* (maps 11, 12, 13 and 16) with the suggestion of a spur-like grouping in *Lupus* (maps 13 and 16). A clear break in the line of the Milky Way caused by a dust cloud is seen half-way through *Vela* (map 16). Other seeming voids in the Milky Way produced by dust clouds are seen close to *Acrux* (map 16) and close to *Deneb* (map 8), with the southern end of the 'Cygnus Rift' well defined in the region west of *Altair* (also map 8).

The same *Cygnus* region provides one of the best examples of an identification group of stars that are entirely disassociated in three-dimensional space, the delta wing grouping in *Cygnus*. The distance given by trigonometric parallax (tr) from *Hipparcos* makes a most interesting comparison with an estimate by spectroscopic parallax (sp): *Alpha,* tr 990 pc (*V* 1.25, A2Ia) sp 520 pc; *Gamma,* tr 470 pc (*V* 2.20, F8Ib) sp 240 pc; *Delta,* tr 52 pc (*V* 2.86, B9.5III) sp 50 pc; *Epsilon*, tr 22 pc (*V* 2.46, K0III) sp 22 pc. The convergence at the near end of the range is striking, and both methods produce the same ranking nearest to farthest, but the different measures to the remote and rare supergiants *Alpha* and *Gamma* are the point of most interest. No weight can be placed on such a small sample, but it may be remarked that spectroscopic parallax is a derived method of estimating distance that ultimately depends on the yardstick of direct measurement, trigonometric parallax. The distances to very remote objects that are found by either method must be treated with great caution.

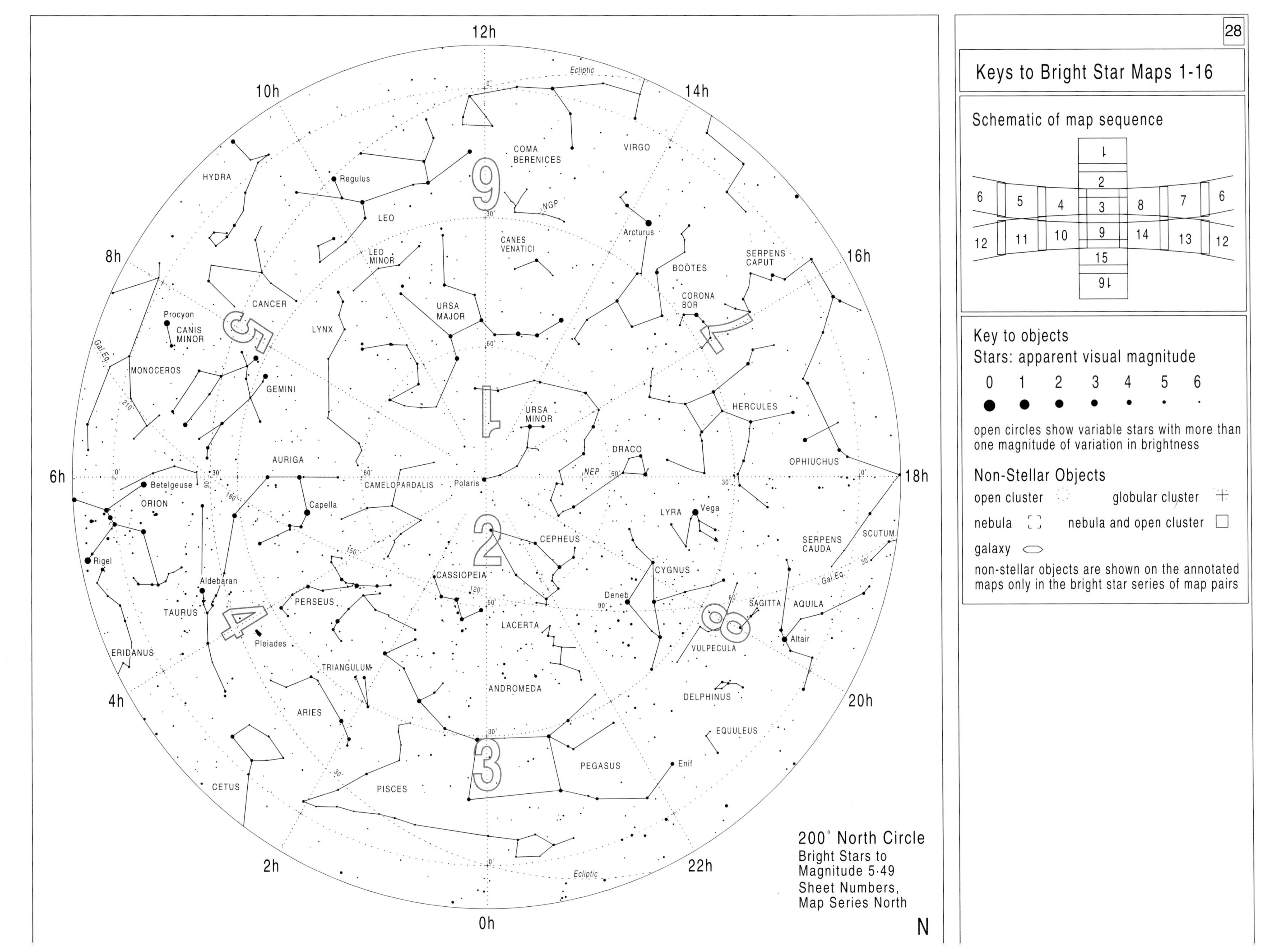
28
Keys to Bright Star Maps 1-16
Schematic of map sequence
1
2
3
9
15
16
6 5 4 8 7 6
12 11 10 14 13 12
Key to objects
Stars: apparent visual magnitude
0 1 2 3 4 5 6
open circles show variable stars with more than one magnitude of variation in brightness
Non-Stellar Objects
open cluster
globular cluster
nebula
nebula and open cluster
galaxy
non-stellar objects are shown on the annotated maps only in the bright star series of map pairs
200° North Circle
Bright Stars to Magnitude 5·49
Sheet Numbers, Map Series North
N
12h
10h
14h
8h
16h
6h
18h
4h
20h
2h
22h
0h
Ecliptic
Gal.Eq.
NGP
NEP
Polaris
COMA BERENICES
VIRGO
HYDRA
Regulus
LEO
Arcturus
LEO MINOR
CANES VENATICI
BOÖTES
SERPENS CAPUT
CORONA BOR
CANCER
Procyon
CANIS MINOR
LYNX
URSA MAJOR
MONOCEROS
GEMINI
URSA MINOR
HERCULES
AURIGA
DRACO
OPHIUCHUS
Betelgeuse
ORION
CAMELOPARDALIS
Capella
Rigel
CEPHEUS
LYRA
Vega
SERPENS CAUDA
SCUTUM
CASSIOPEIA
CYGNUS
Aldebaran
TAURUS
PERSEUS
Deneb
SAGITTA
AQUILA
Pleiades
LACERTA
Altair
VULPECULA
ERIDANUS
TRIANGULUM
ANDROMEDA
DELPHINUS
ARIES
EQUULEUS
PEGASUS
Enif
CETUS
PISCES

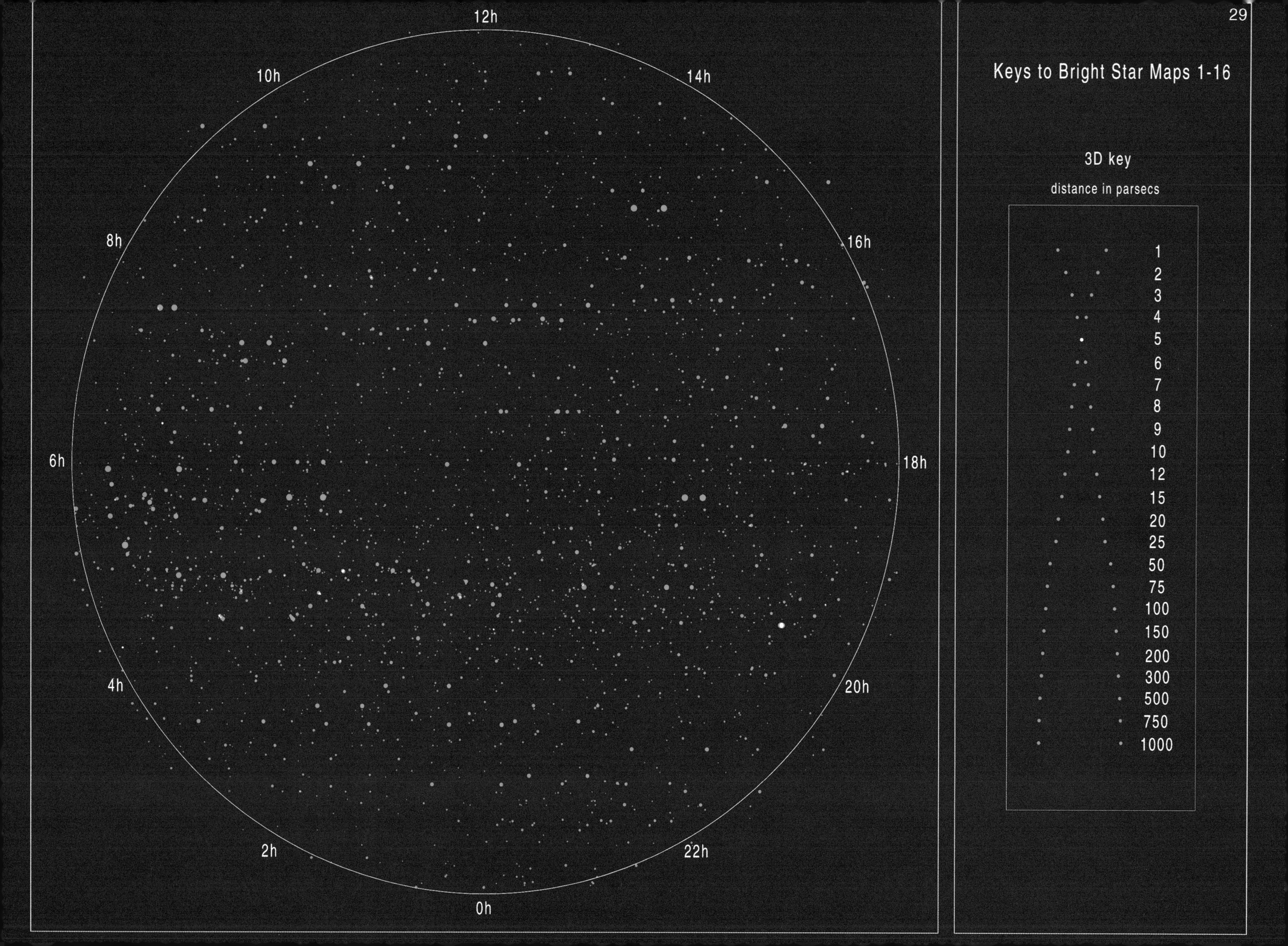
12h
10h
14h
8h
16h
6h
18h
4h
20h
2h
22h
0h
Keys to Bright Star Maps 1-16
3D key
distance in parsecs
1
2
3
4
5
6
7
8
9
10
12
15
20
25
50
75
100
150
200
300
500
750
1000

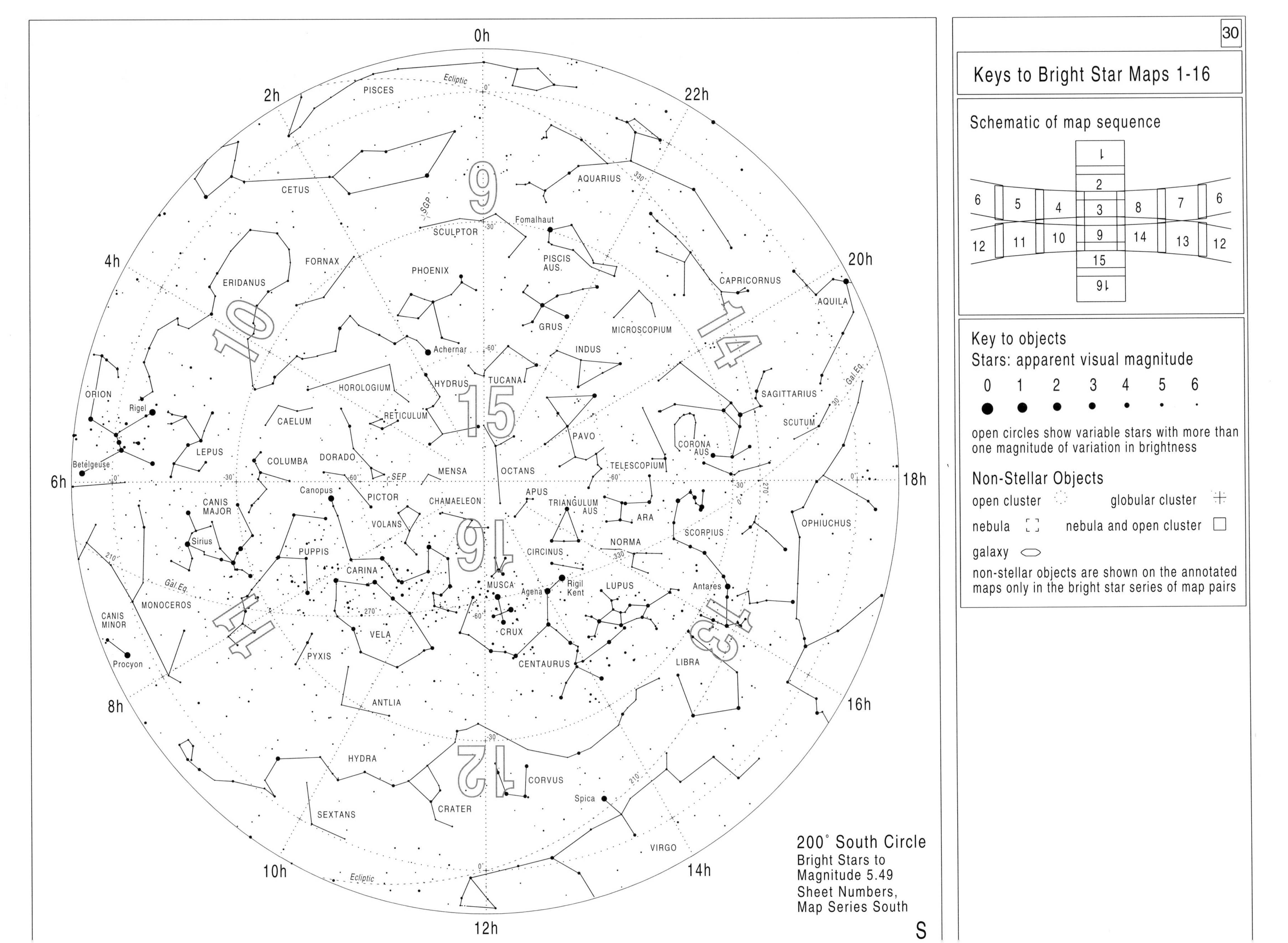

0h
2h
4h
6h
8h
10h
12h
14h
16h
18h
20h
22h
Ecliptic
Gal.Eq.
SGP
SEP
PISCES
CETUS
AQUARIUS
SCULPTOR
Fomalhaut
PISCIS AUS.
FORNAX
ERIDANUS
PHOENIX
GRUS
MICROSCOPIUM
CAPRICORNUS
AQUILA
INDUS
Achernar
HYDRUS
TUCANA
HOROLOGIUM
RETICULUM
CAELUM
ORION
Rigel
Betelgeuse
LEPUS
COLUMBA
DORADO
MENSA
OCTANS
PAVO
SAGITTARIUS
SCUTUM
CORONA AUS
TELESCOPIUM
CANIS MAJOR
Sirius
Canopus
PICTOR
CHAMAELEON
VOLANS
APUS
TRIANGULUM AUS
ARA
OPHIUCHUS
SCORPIUS
NORMA
CIRCINUS
PUPPIS
CARINA
MUSCA
Agena
Rigil Kent
LUPUS
Antares
MONOCEROS
CANIS MINOR
Procyon
VELA
CRUX
PYXIS
CENTAURUS
LIBRA
ANTLIA
HYDRA
CORVUS
Spica
SEXTANS
CRATER
VIRGO
200° South Circle
Bright Stars to
Magnitude 5.49
Sheet Numbers,
Map Series South
S
Keys to Bright Star Maps 1-16
Schematic of map sequence
Key to objects
Stars: apparent visual magnitude
0 1 2 3 4 5 6
open circles show variable stars with more than one magnitude of variation in brightness
Non-Stellar Objects
open cluster
globular cluster
nebula
nebula and open cluster
galaxy
non-stellar objects are shown on the annotated maps only in the bright star series of map pairs

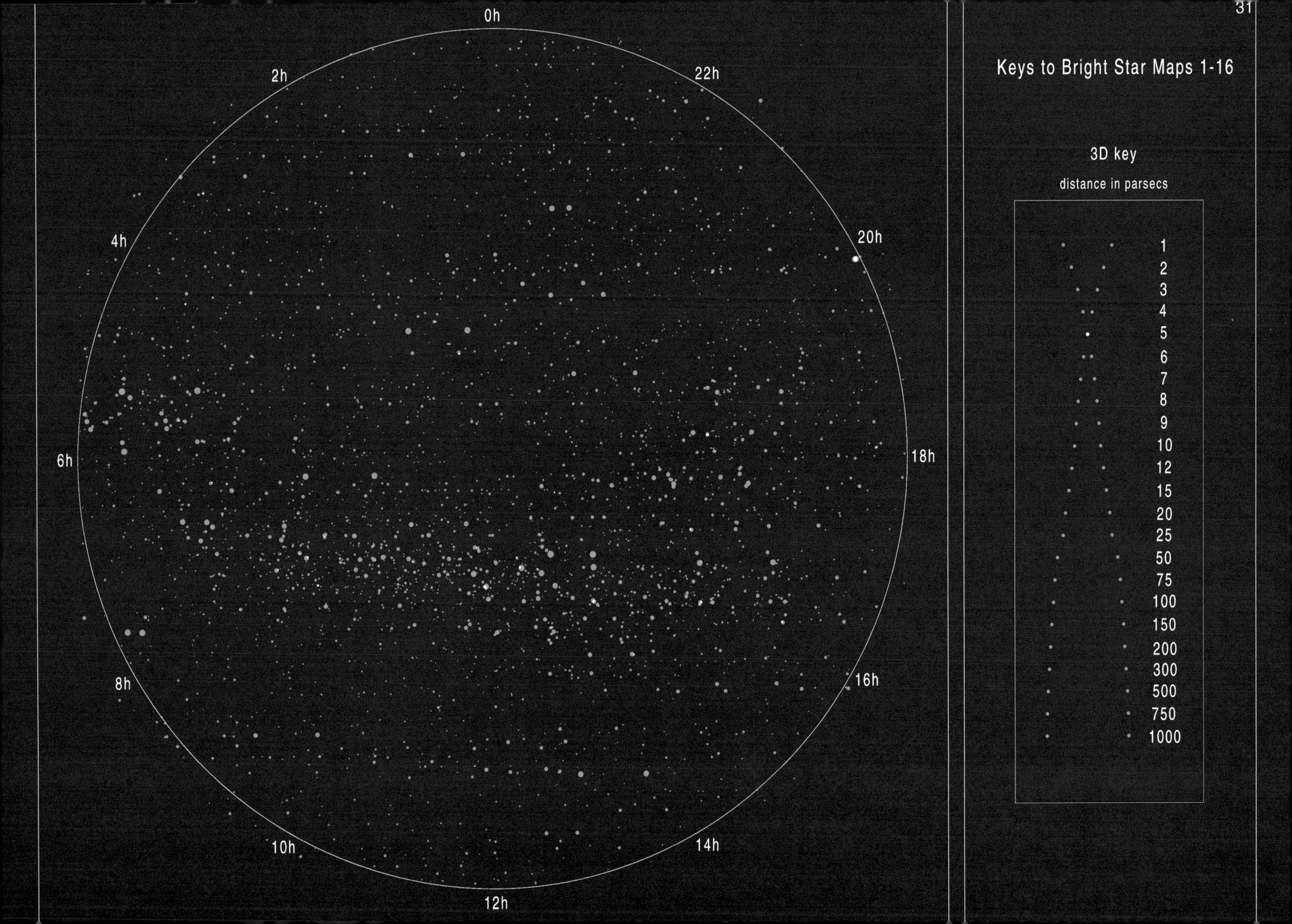
0h
2h
22h
4h
20h
6h
18h
8h
16h
10h
14h
12h
Keys to Bright Star Maps 1-16
3D key
distance in parsecs
1
2
3
4
5
6
7
8
9
10
12
15
20
25
50
75
100
150
200
300
500
750
1000

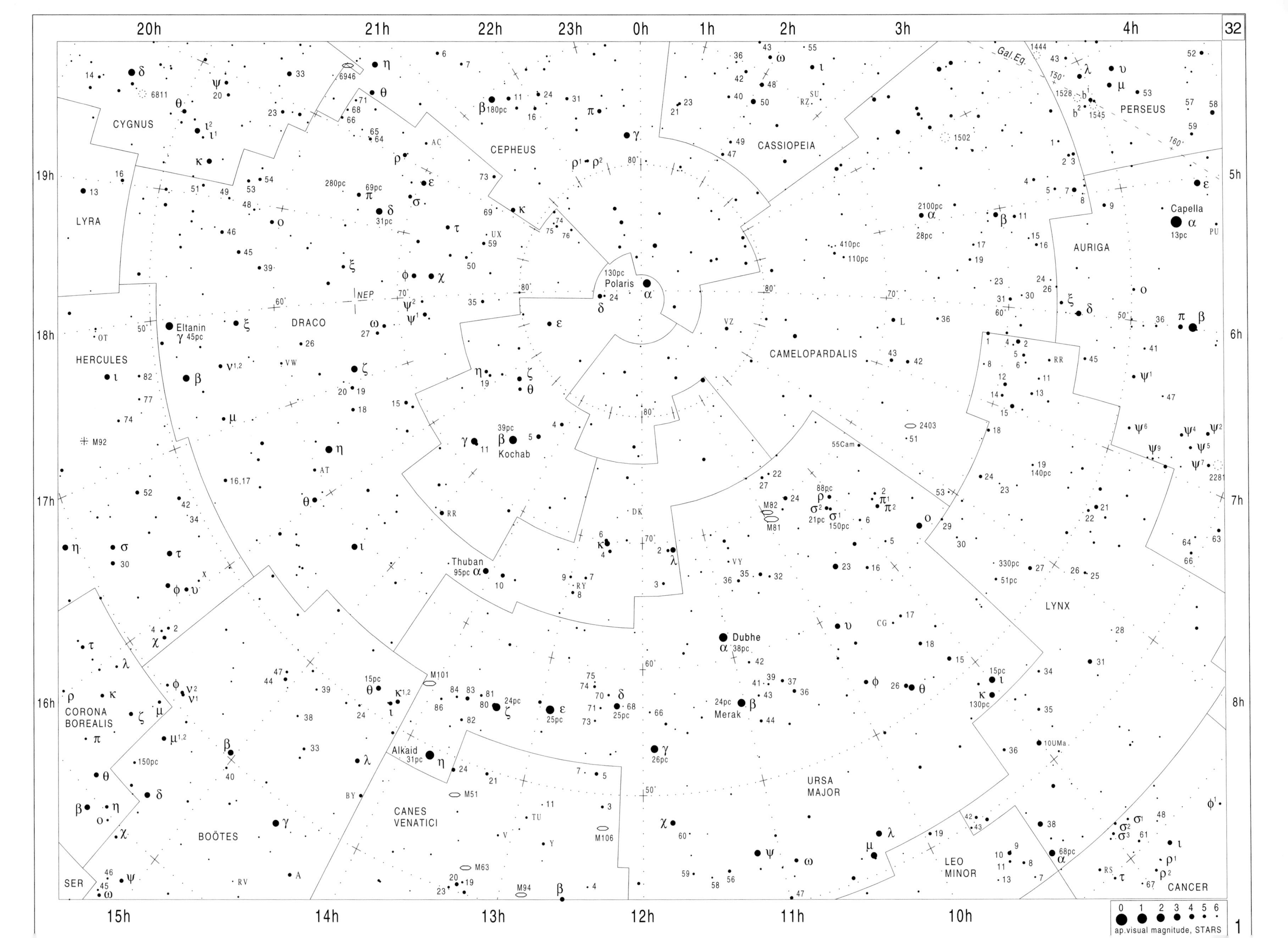
CYGNUS
LYRA
HERCULES
DRACO
CEPHEUS
CASSIOPEIA
PERSEUS
AURIGA
CAMELOPARDALIS
LYNX
URSA MAJOR
CANES VENATICI
BOÖTES
CORONA BOREALIS
SER
LEO MINOR
CANCER
Capella
Polaris
Eltanin
Kochab
Thuban
Dubhe
Merak
Alkaid
NEP
Gal.Eq.
M92
M101
M51
M63
M94
M106
M81
M82
2403
6946
6811
1502
1528
1545
1444
2281
20h
21h
22h
23h
0h
1h
2h
3h
4h
5h
6h
7h
8h
10h
11h
12h
13h
14h
15h
16h
17h
18h
19h
0 1 2 3 4 5 6
ap.visual magnitude, STARS

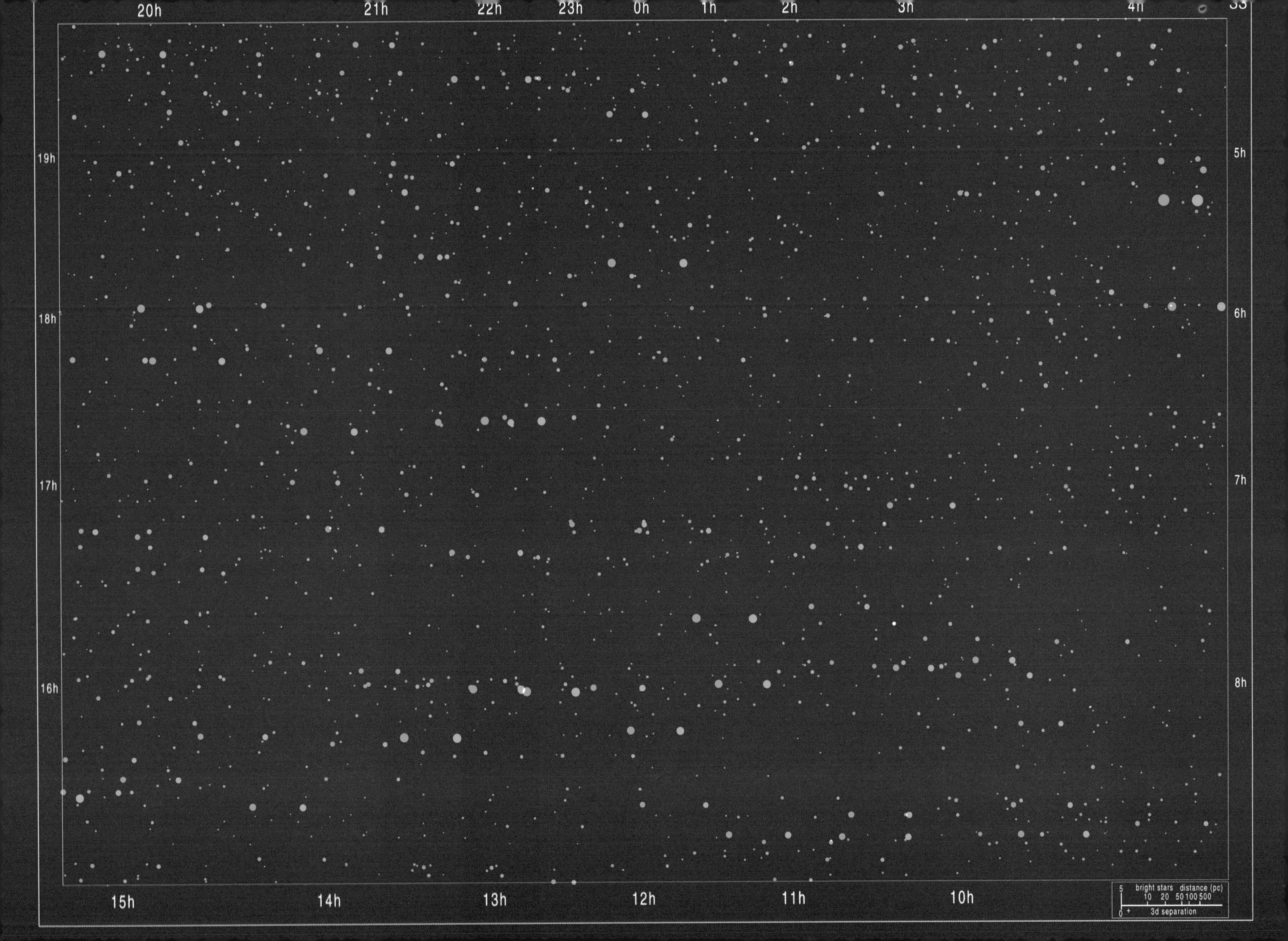

20h
21h
22h
23h
0h
1h
2h
3h
4h
19h
18h
17h
16h
5h
6h
7h
8h
15h
14h
13h
12h
11h
10h
bright stars distance (pc)
5 10 20 50 100 500
0 + 3d separation

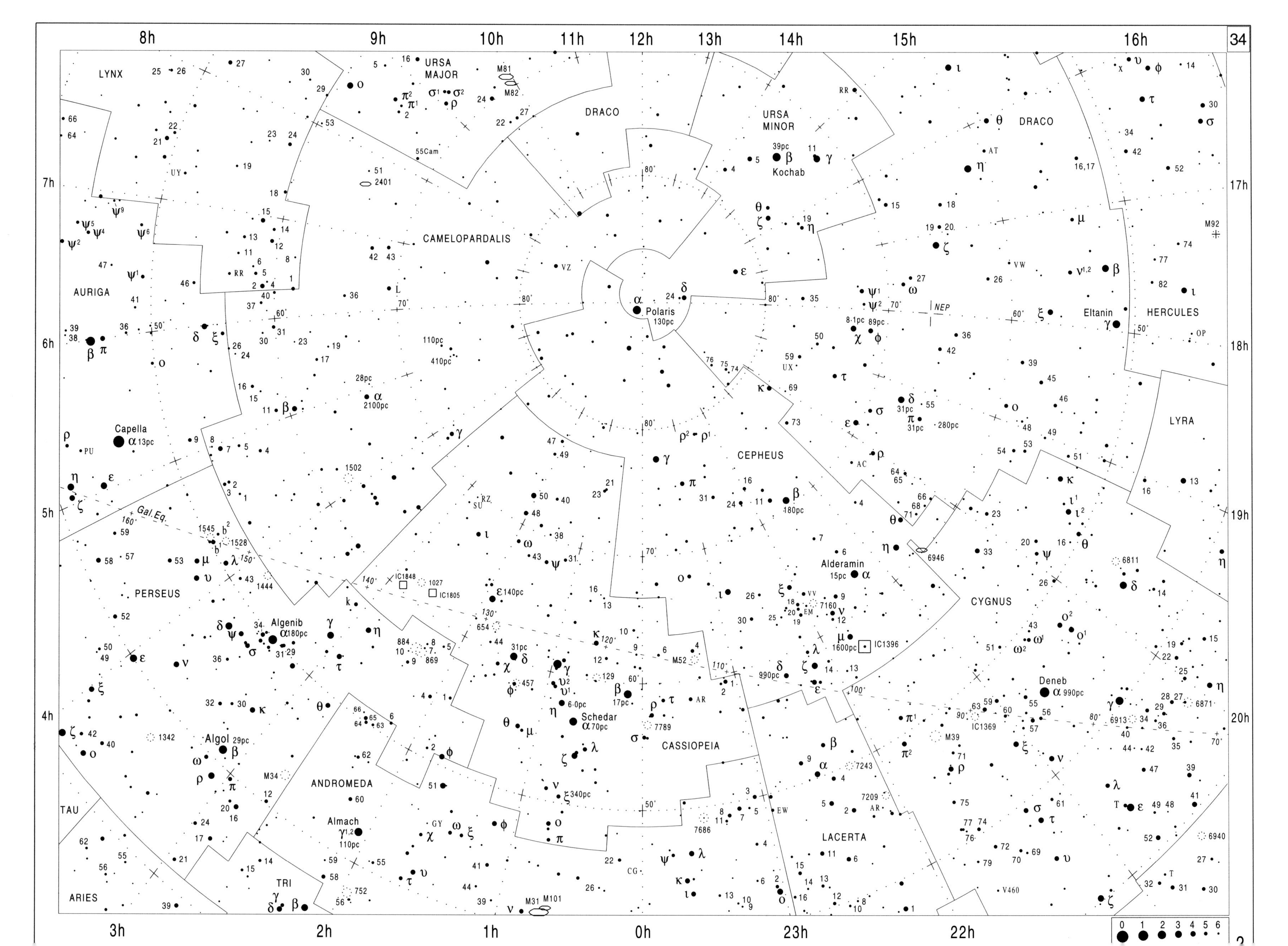

LYNX
URSA MAJOR
DRACO
URSA MINOR
Kochab
DRACO
HERCULES
Eltanin
NEP
CAMELOPARDALIS
AURIGA
Polaris 130pc
LYRA
Capella
CEPHEUS
Alderamin
PERSEUS
Algenib
CYGNUS
Deneb
Schedar
CASSIOPEIA
Algol
ANDROMEDA
Almach
LACERTA
TAU
TRI
ARIES
Gal.Eq.
M81
M82
M92
M52
M39
M34
M31
M101
IC1396
IC1369
IC1805
IC1848
8h
9h
10h
11h
12h
13h
14h
15h
16h
17h
18h
19h
20h
7h
6h
5h
4h
3h
2h
1h
0h
23h
22h

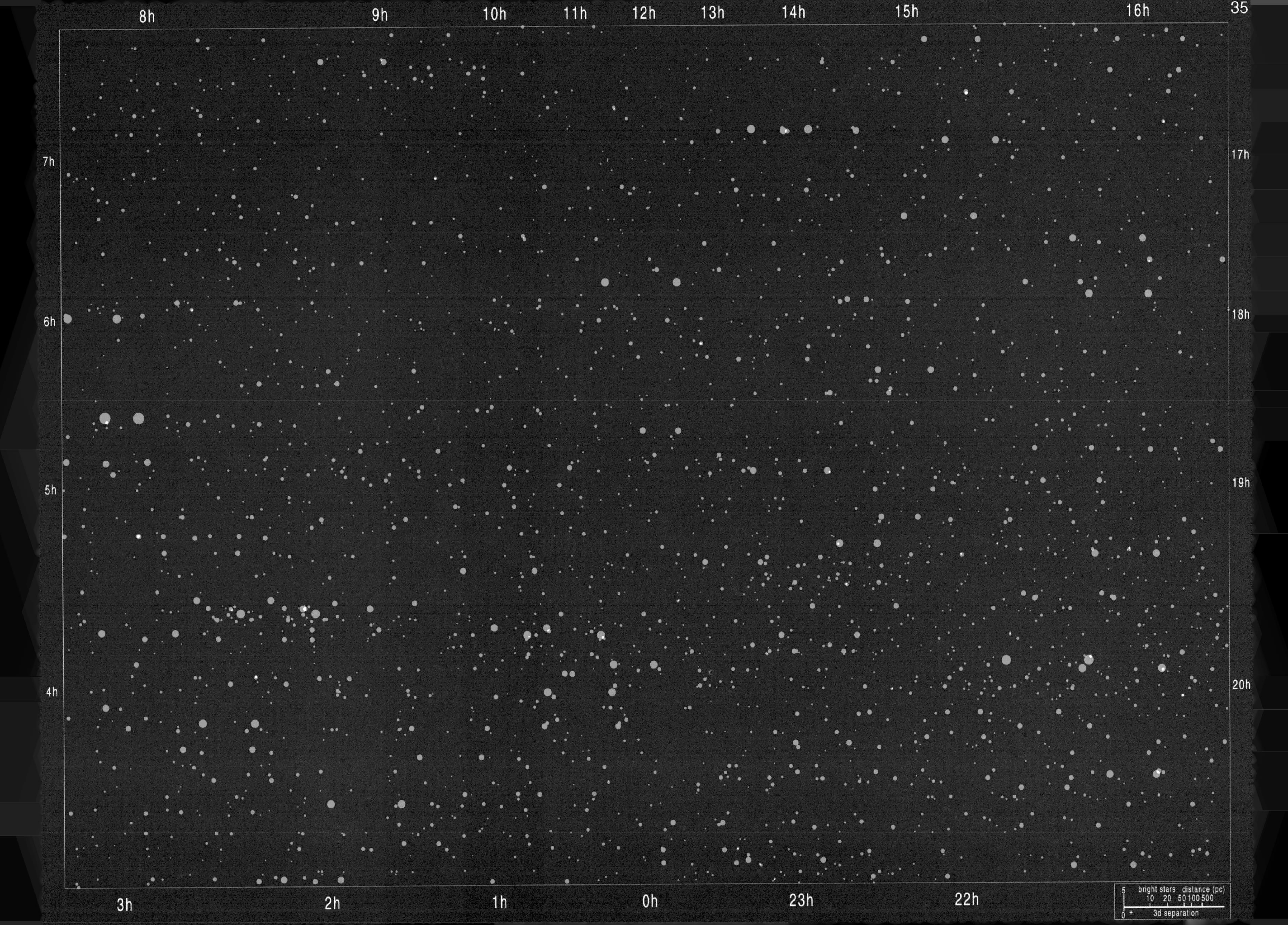

8h
9h
10h
11h
12h
13h
14h
15h
16h
35
7h
6h
5h
4h
17h
18h
19h
20h
3h
2h
1h
0h
23h
22h
bright stars
distance (pc)
5
10 20 50 100 500
0
3d separation

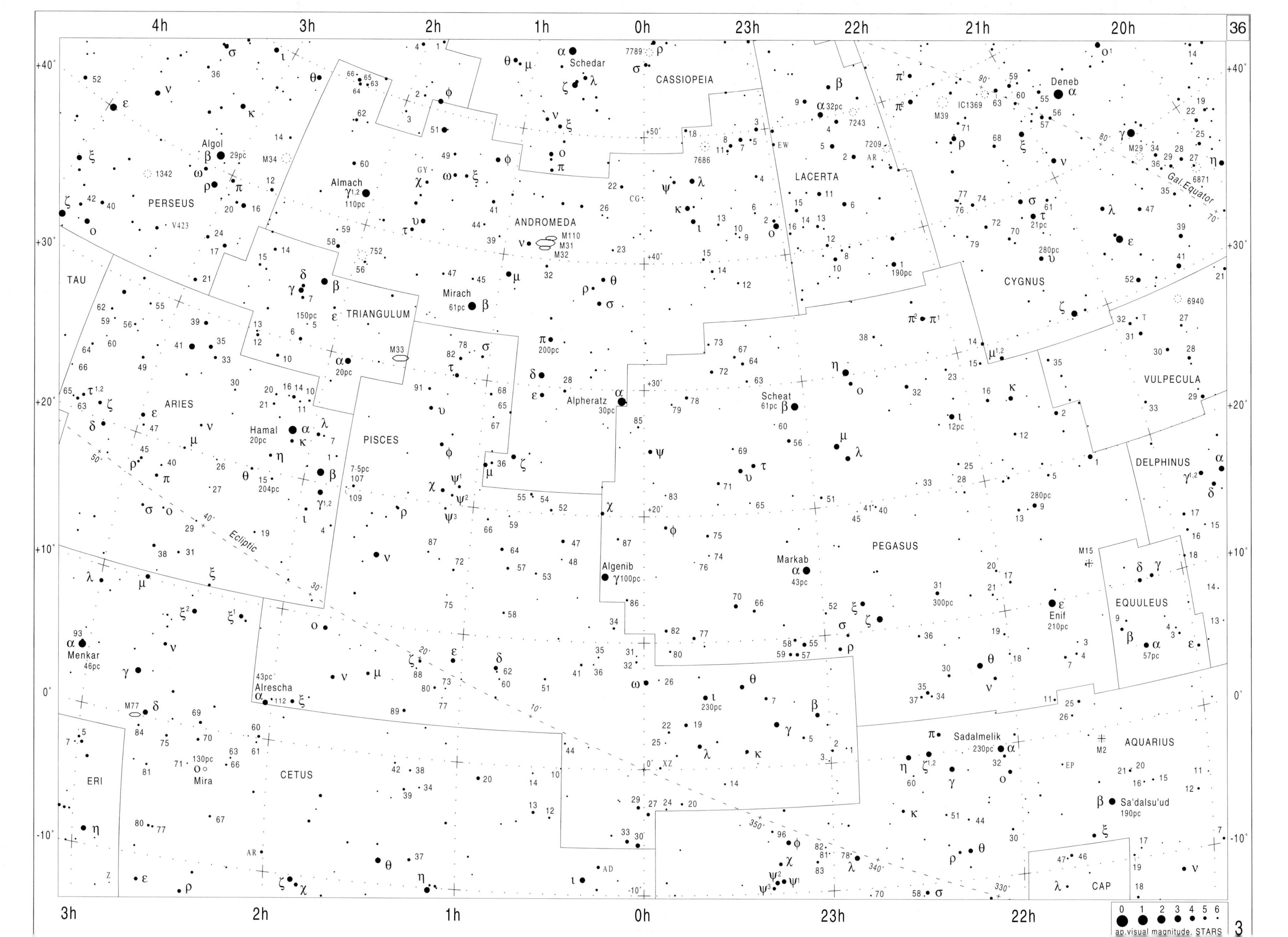
4h
3h
2h
1h
0h
23h
22h
21h
20h
+40°
+30°
+20°
+10°
0°
-10°
CASSIOPEIA
Schedar
LACERTA
CYGNUS
Deneb
Gal. Equator
PERSEUS
Algol
ANDROMEDA
Almach
Mirach
Alpheratz
TRIANGULUM
TAU
ARIES
Hamal
PISCES
VULPECULA
DELPHINUS
PEGASUS
Scheat
Markab
Algenib
Enif
EQUULEUS
Ecliptic
CETUS
Menkar
Alrescha
Mira
ERI
AQUARIUS
Sadalmelik
Sa'dalsu'ud
CAP
M31
M32
M110
M33
M34
M77
M15
M2
M29
M39
IC1369
0 1 2 3 4 5 6
ap. visual magnitude, STARS

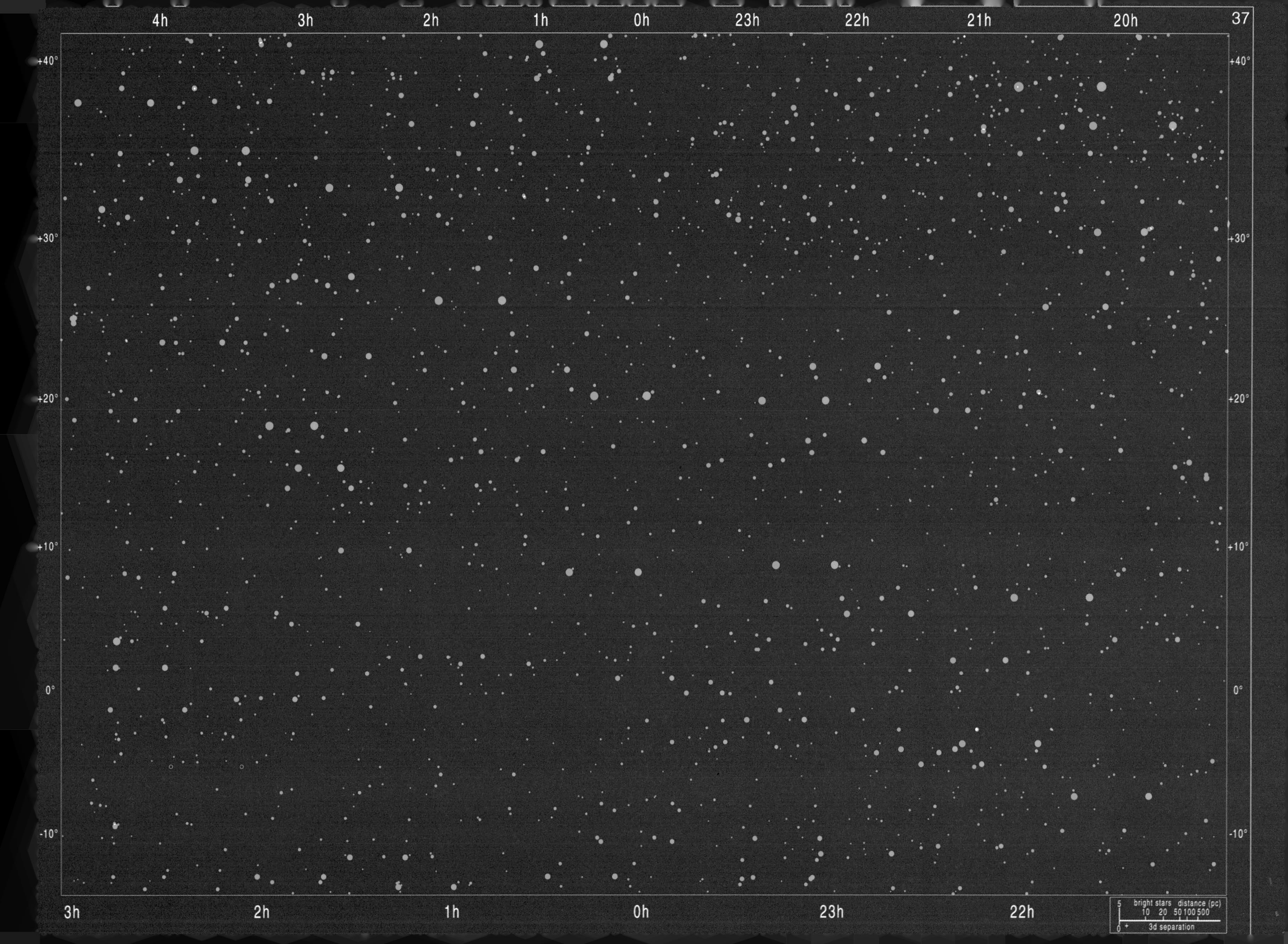

4h
3h
2h
1h
0h
23h
22h
21h
20h
+40°
+30°
+20°
+10°
0°
-10°
3h
2h
1h
0h
23h
22h
bright stars distance (pc)
5
10 20 50 100 500
0
3d separation

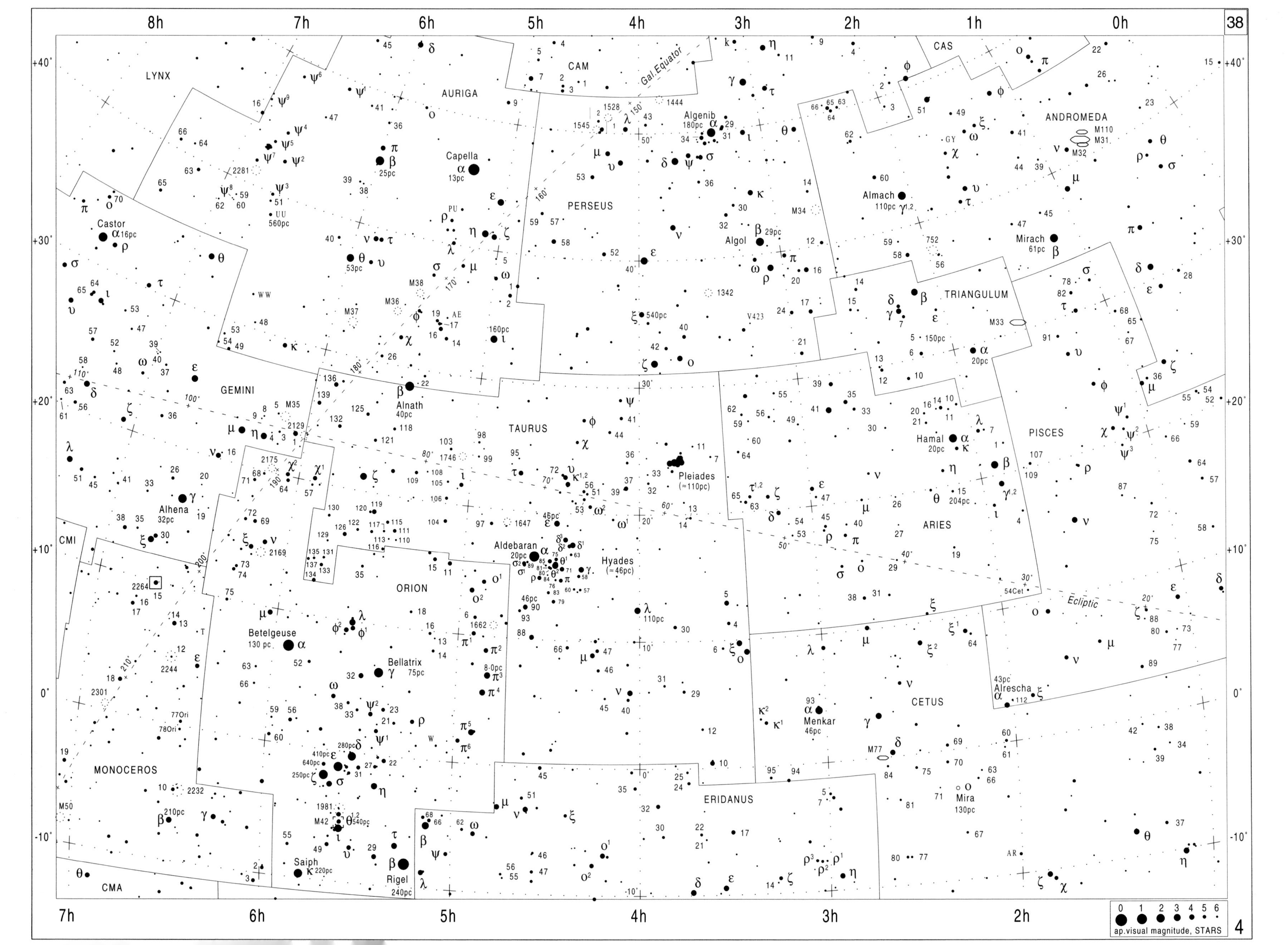

LYNX
AURIGA
CAM
CAS
ANDROMEDA
PERSEUS
TRIANGULUM
PISCES
ARIES
TAURUS
GEMINI
ORION
CETUS
ERIDANUS
MONOCEROS
CMI
CMA
Gal. Equator
Ecliptic
Capella
13pc
Castor
16pc
Alnath
40pc
Algenib
180pc
Algol
29pc
Almach
110pc
Mirach
61pc
Hamal
20pc
Pleiades
(≈110pc)
Hyades
(≈46pc)
Aldebaran
20pc
Alhena
32pc
Betelgeuse
130 pc
Bellatrix
75pc
Rigel
240pc
Saiph
220pc
Menkar
46pc
Mira
130pc
Alrescha
43pc
M31
M32
M110
M33
M34
M35
M36
M37
M38
M42
M50
M77
0 1 2 3 4 5 6
ap. visual magnitude, STARS

8h
7h
6h
5h
4h
3h
2h
1h
0h
+40°
+30°
+20°
+10°
0°
-10°
7h
6h
5h
4h
3h
2h
5
0
bright stars distance (pc)
10 20 50 100 500
3d separation

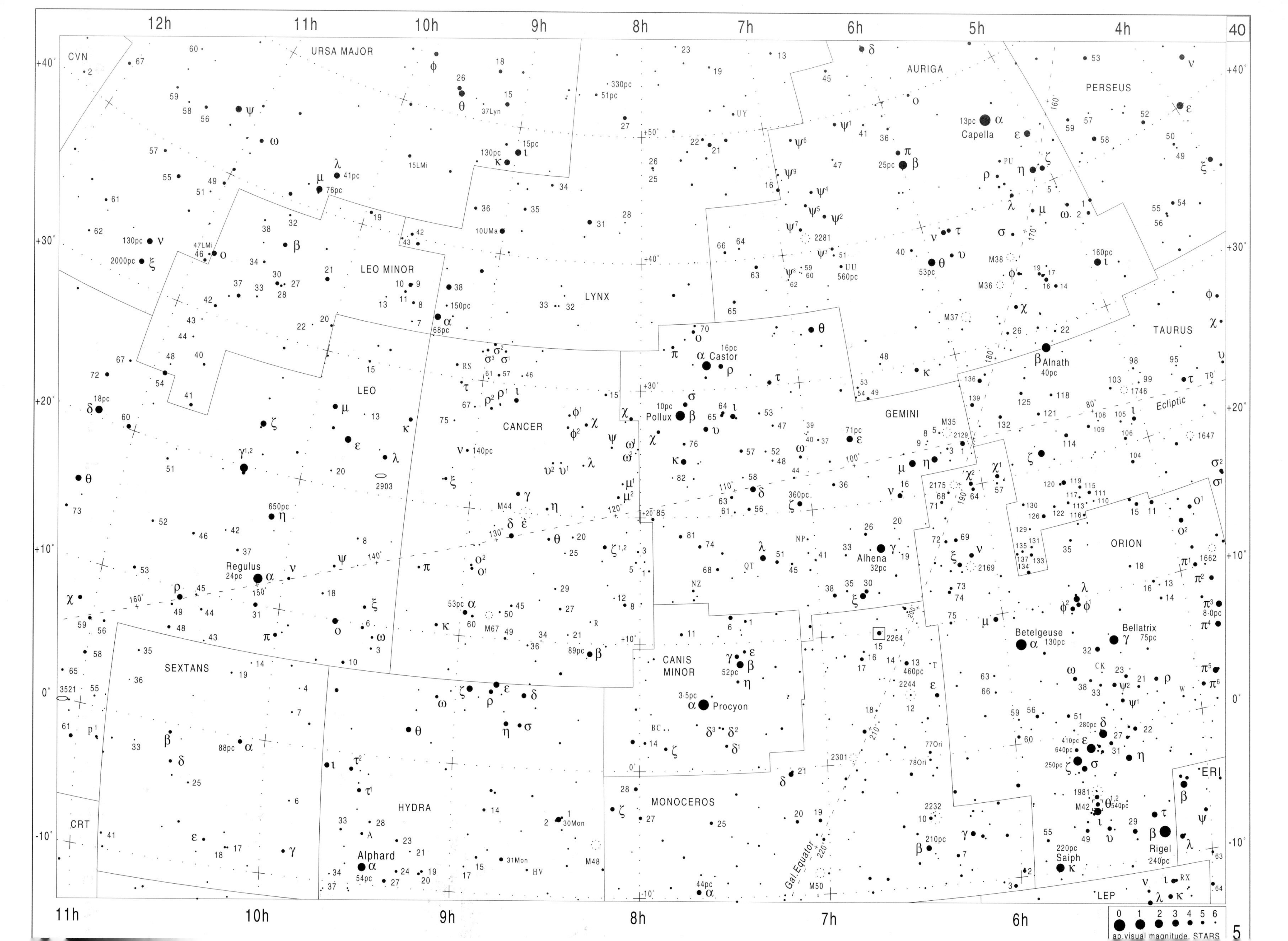
URSA MAJOR
AURIGA
PERSEUS
CVN
LEO MINOR
LYNX
TAURUS
LEO
CANCER
GEMINI
ORION
SEXTANS
CANIS MINOR
MONOCEROS
HYDRA
CRT
ERI
LEP
Capella
Alnath
Castor
Pollux
Regulus
Alhena
Betelgeuse
Bellatrix
Procyon
Alphard
Rigel
Saiph
Ecliptic
Gal.Equator
ap.visual magnitude STARS

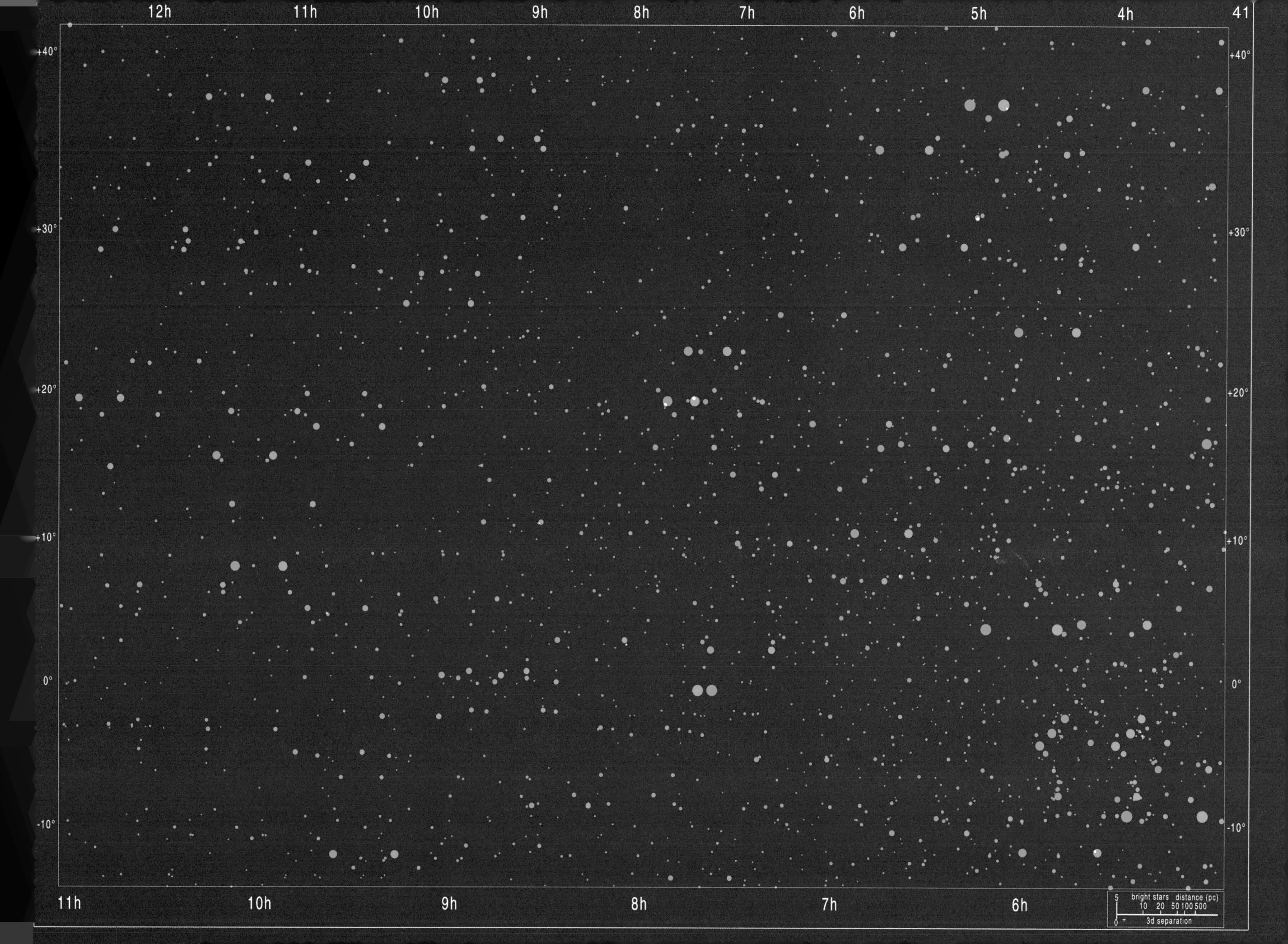

12h
11h
10h
9h
8h
7h
6h
5h
4h
41
+40°
+30°
+20°
+10°
0°
-10°
11h
10h
9h
8h
7h
6h
bright stars distance (pc)
5 10 20 50 100 500
0 +
3d separation

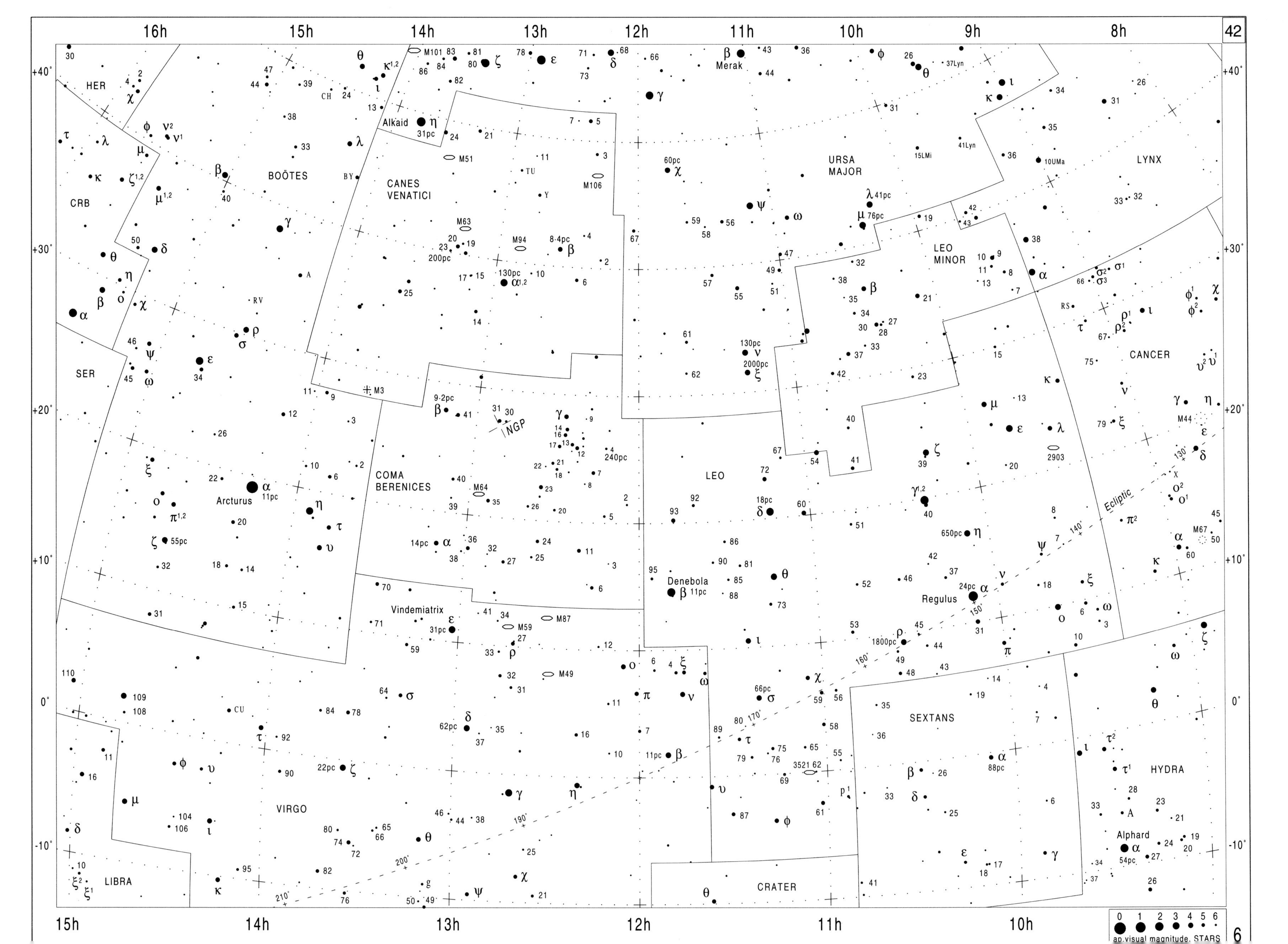
16h
15h
14h
13h
12h
11h
10h
9h
8h
+40°
+30°
+20°
+10°
0°
-10°
HER
CRB
SER
BOÖTES
CANES
VENATICI
URSA
MAJOR
LYNX
LEO
MINOR
CANCER
COMA
BERENICES
LEO
VIRGO
SEXTANS
HYDRA
CRATER
LIBRA
Alkaid
Merak
Arcturus
Denebola
Regulus
Vindemiatrix
Alphard
Ecliptic
NGP
M101
M51
M106
M63
M94
M3
M64
M87
M59
M49
M44
M67
2903
3521
0 1 2 3 4 5 6
ap. visual magnitude, STARS

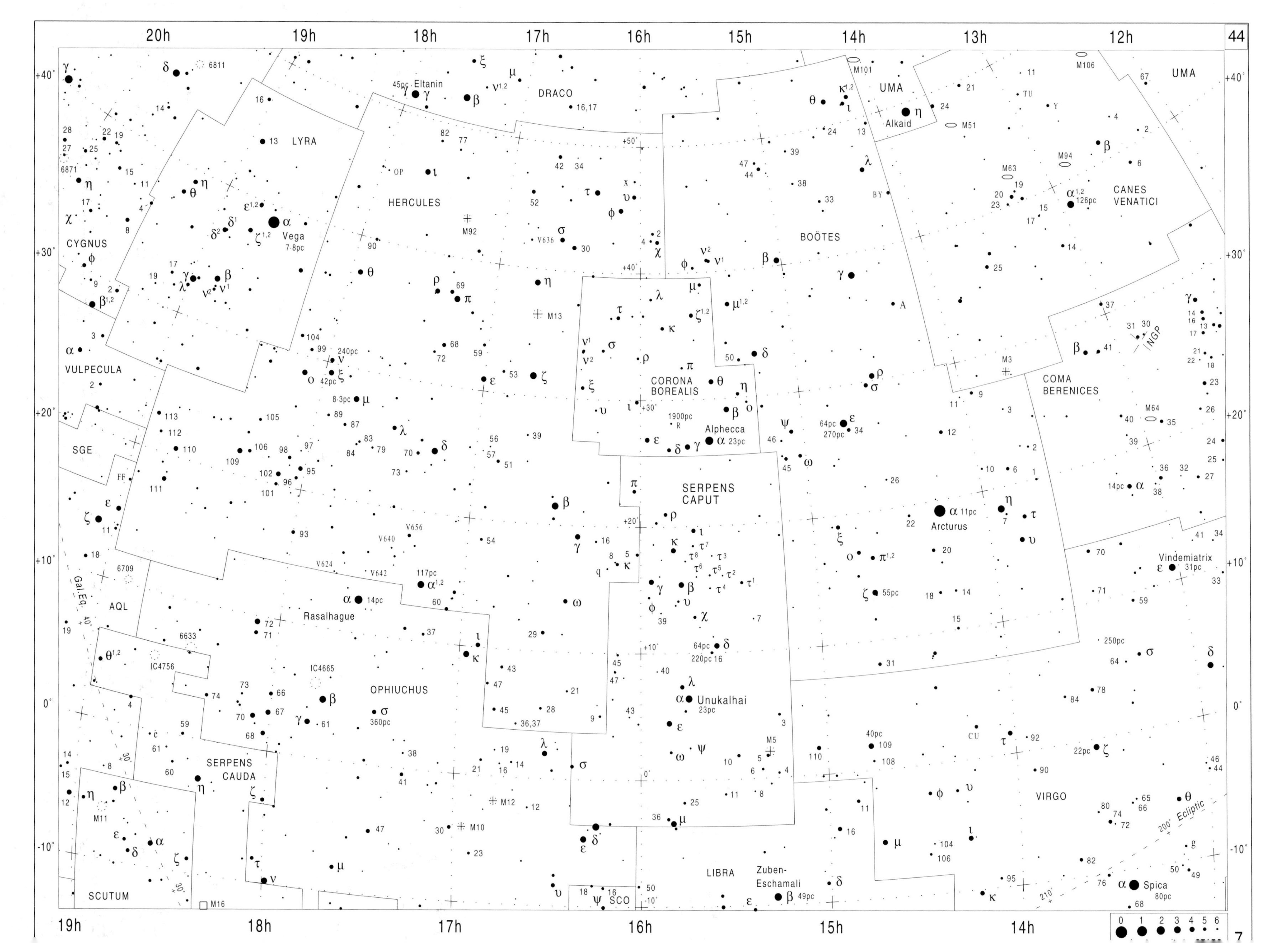

44
7
12h
13h
14h
15h
16h
17h
18h
19h
20h
+40°
+30°
+20°
+10°
0°
-10°
UMA
CANES VENATICI
COMA BERENICES
NGP
Vindemiatrix
VIRGO
Ecliptic
Spica
Alkaid
BOÖTES
Arcturus
CORONA BOREALIS
Alphecca
SERPENS CAPUT
Unukalhai
LIBRA
Zuben-Eschamali
SCO
DRACO
HERCULES
Eltanin
OPHIUCHUS
Rasalhague
LYRA
Vega
CYGNUS
VULPECULA
SGE
AQL
SERPENS CAUDA
SCUTUM
Gal.Eq.
M106
M94
M63
M51
M101
M3
M64
M5
M13
M92
M12
M10
M16
M11
IC4665
IC4756
6633
6709
6811
6871

20h
19h
18h
17h
16h
15h
14h
13h
12h
45
+40°
+30°
+20°
+10°
0°
-10°
19h
18h
17h
16h
15h
14h
bright stars
distance (pc)
5
10 20 50 100 500
0
3d separation

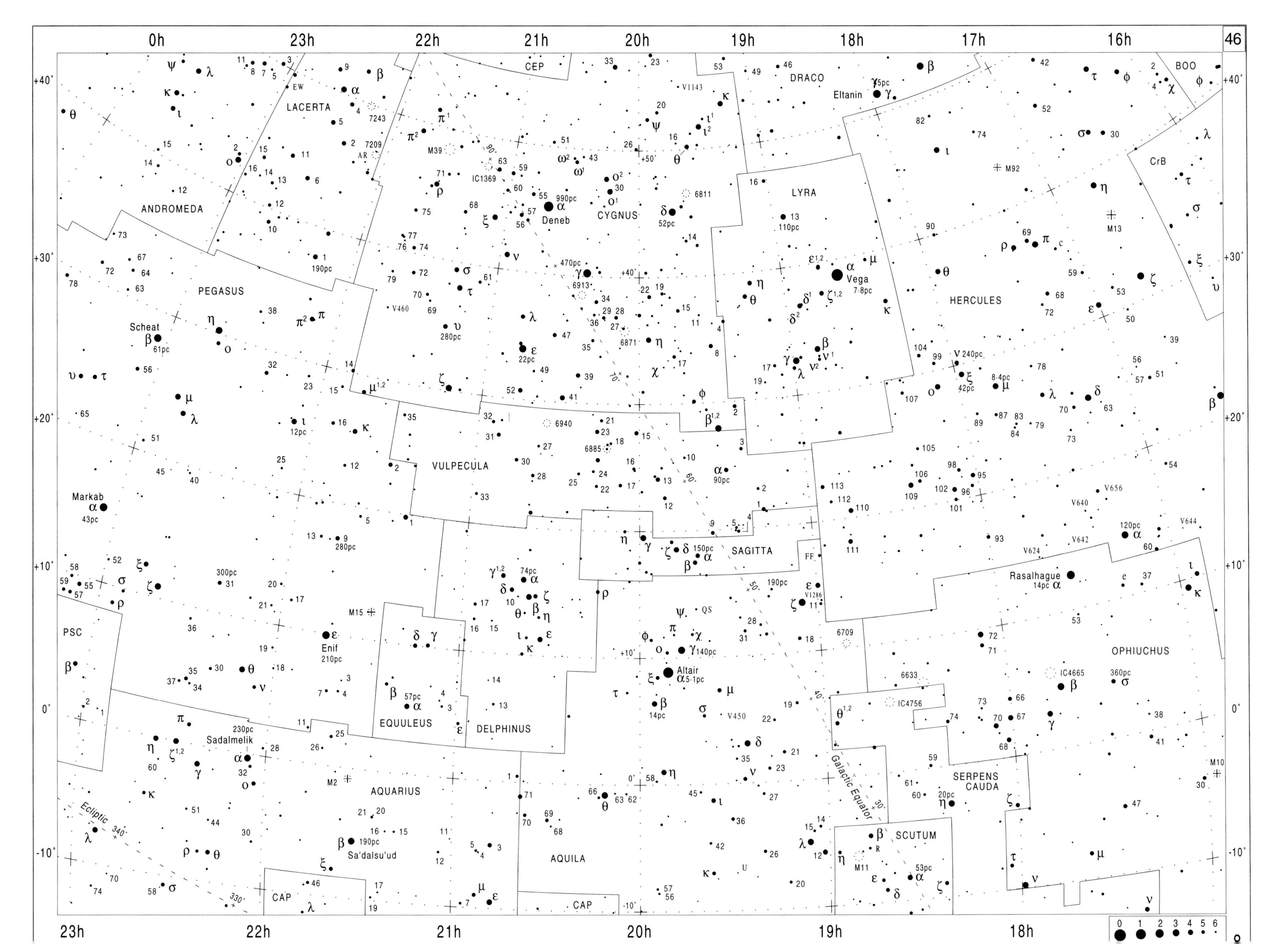
0h
23h
22h
21h
20h
19h
18h
17h
16h
+40°
+30°
+20°
+10°
0°
-10°
BOO
CEP
DRACO
Eltanin
LACERTA
CrB
M92
M13
M39
IC1369
Deneb
990pc
CYGNUS
6811
52pc
LYRA
110pc
Vega
7·8pc
HERCULES
240pc
42pc
8·4pc
ANDROMEDA
PEGASUS
Scheat
61pc
190pc
280pc
22pc
470pc
6913
6871
6940
6885
VULPECULA
90pc
Markab
43pc
SAGITTA
150pc
FF
V1286
190pc
120pc
V644
V642
V624
V640
V656
Rasalhague
14pc
OPHIUCHUS
IC4665
360pc
M15
74pc
PSC
Enif
210pc
300pc
EQUULEUS
57pc
DELPHINUS
Altair
5·1pc
140pc
14pc
QS
V450
6709
6633
IC4756
SERPENS
CAUDA
20pc
M10
Sadalmelik
230pc
M2
AQUARIUS
Sa'dalsu'ud
AQUILA
SCUTUM
53pc
M11
Galactic Equator
Ecliptic
CAP
V1143
V460
7243
7209
AR
EW
0 1 2 3 4 5 6

47
0h
23h
22h
21h
20h
19h
18h
17h
16h
+40°
+30°
+20°
+10°
0°
-10°
23h
22h
21h
20h
19h
18h
bright stars
distance (pc)
5
10 20 50 100 500
0
3d separation

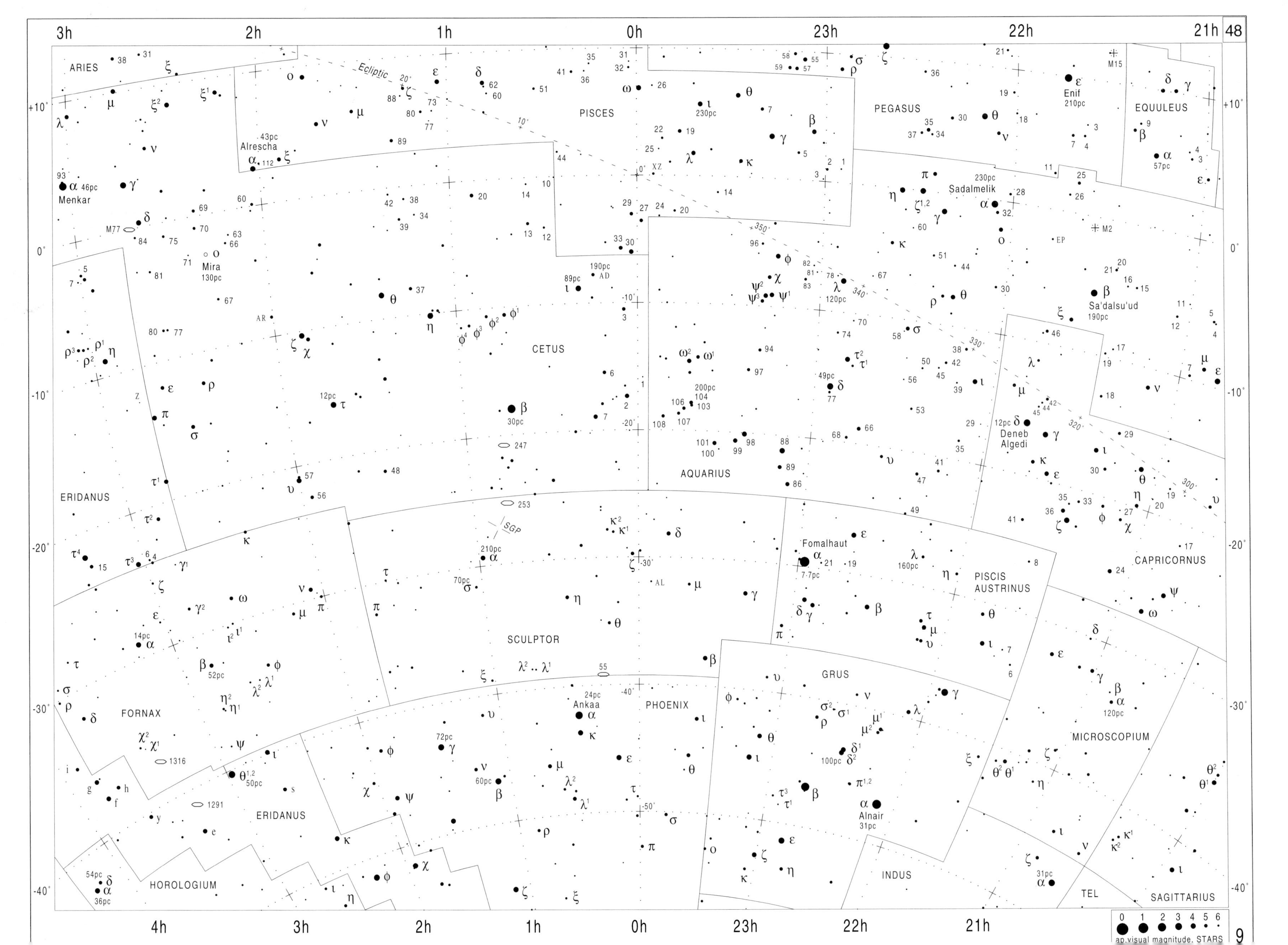

3h
2h
1h
0h
23h
22h
21h
+10°
0°
-10°
-20°
-30°
-40°
4h
ARIES
PISCES
PEGASUS
EQUULEUS
CETUS
AQUARIUS
CAPRICORNUS
PISCIS AUSTRINUS
ERIDANUS
SCULPTOR
FORNAX
PHOENIX
GRUS
MICROSCOPIUM
INDUS
TEL
SAGITTARIUS
HOROLOGIUM
Ecliptic
SGP
Menkar
46pc
Mira
130pc
Alrescha
43pc
Enif
210pc
Sadalmelik
230pc
Sa'dalsu'ud
190pc
Deneb Algedi
12pc
Fomalhaut
7.7pc
Ankaa
24pc
Alnair
31pc
M77
M15
M2
0 1 2 3 4 5 6
ap.visual magnitude, STARS

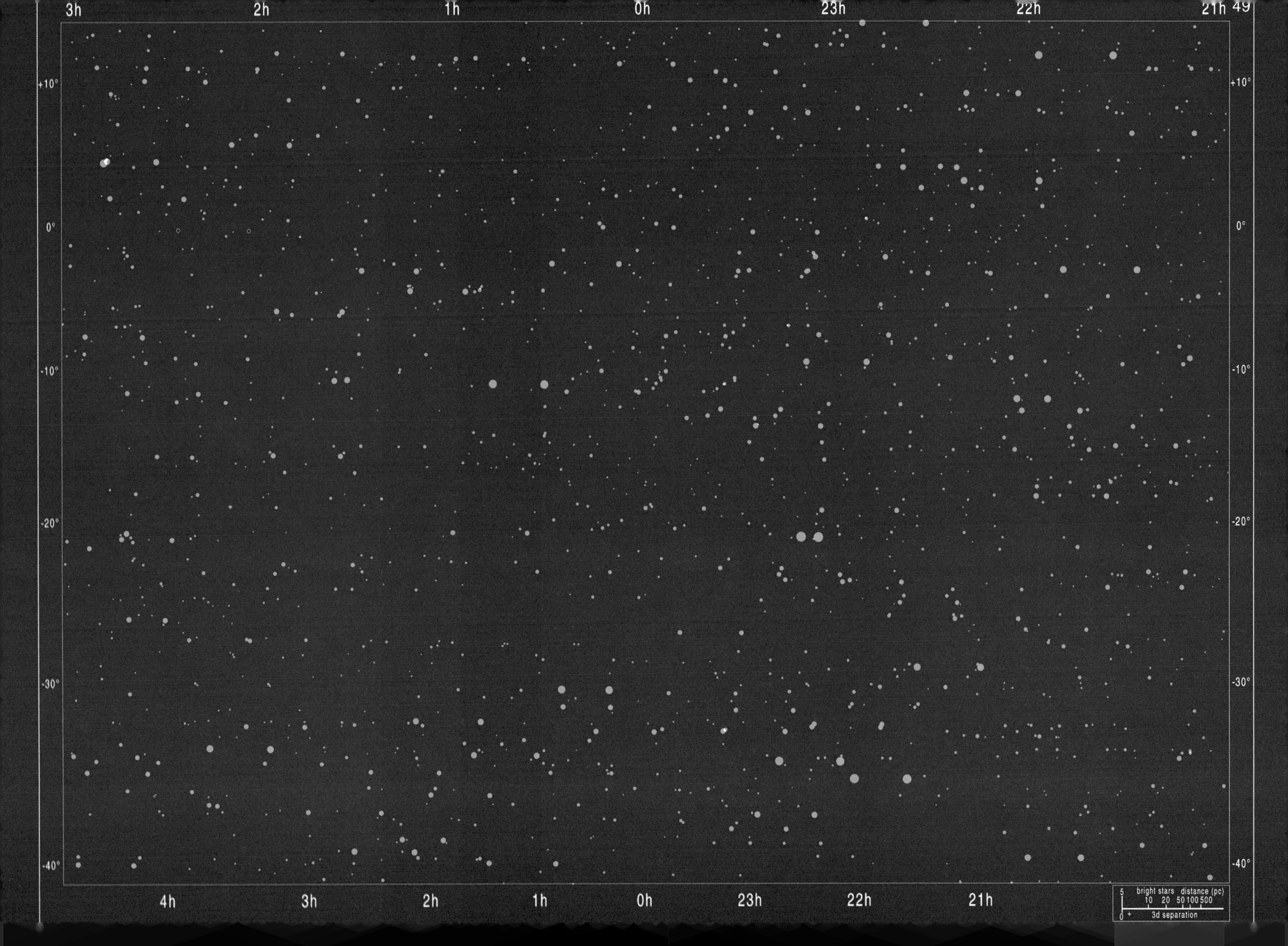

3h
2h
1h
0h
23h
22h
21h 49
+10°
0°
-10°
-20°
-30°
-40°
4h
3h
2h
1h
0h
23h
22h
21h
bright stars distance (pc)
5
10 20 50 100 500
0
3d separation

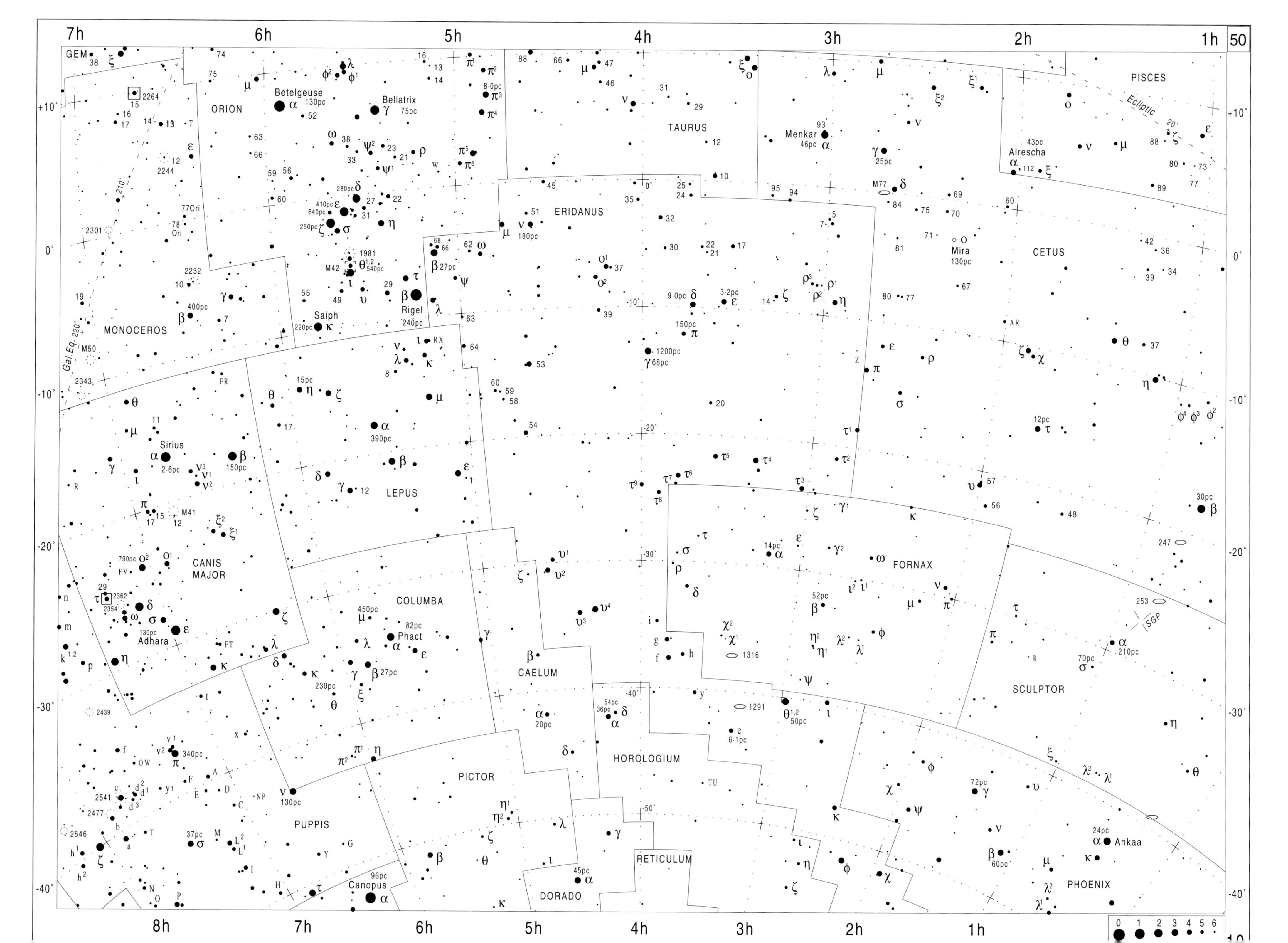

50
PISCES
Ecliptic
CETUS
SCULPTOR
PHOENIX
FORNAX
TAURUS
ERIDANUS
HOROLOGIUM
RETICULUM
CAELUM
DORADO
PICTOR
LEPUS
COLUMBA
ORION
MONOCEROS
CANIS MAJOR
PUPPIS
GEM
Alrescha
Menkar
Mira
Ankaa
Bellatrix
Betelgeuse
Rigel
Saiph
Sirius
Adhara
Phact
Canopus
SGP
Gal.Eq.
7h
6h
5h
4h
3h
2h
1h
8h

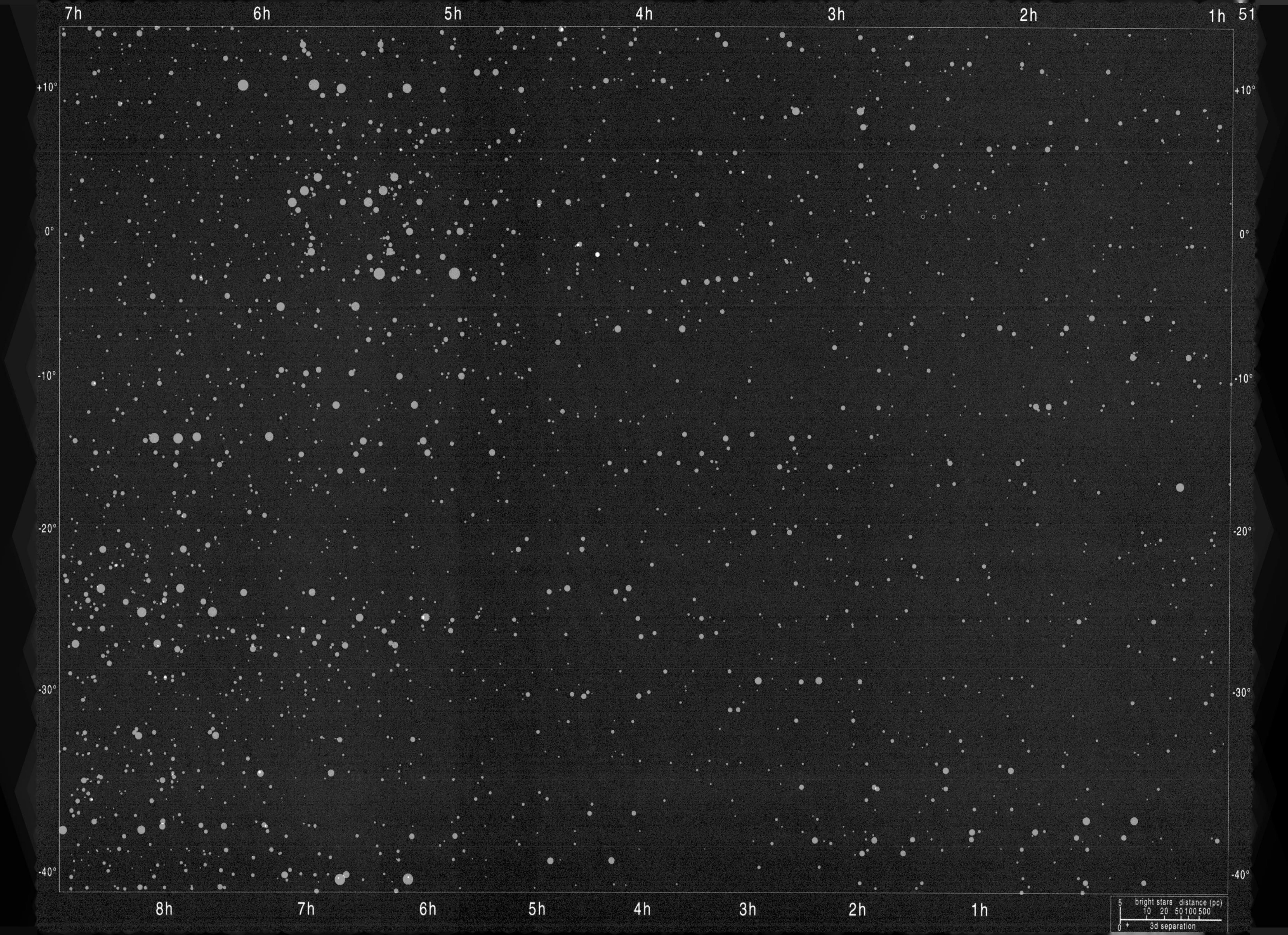
7h
6h
5h
4h
3h
2h
1h
51
+10°
0°
-10°
-20°
-30°
-40°
8h
7h
6h
5h
4h
3h
2h
1h
bright stars
distance (pc)
5
10 20 50 100 500
0
3d separation

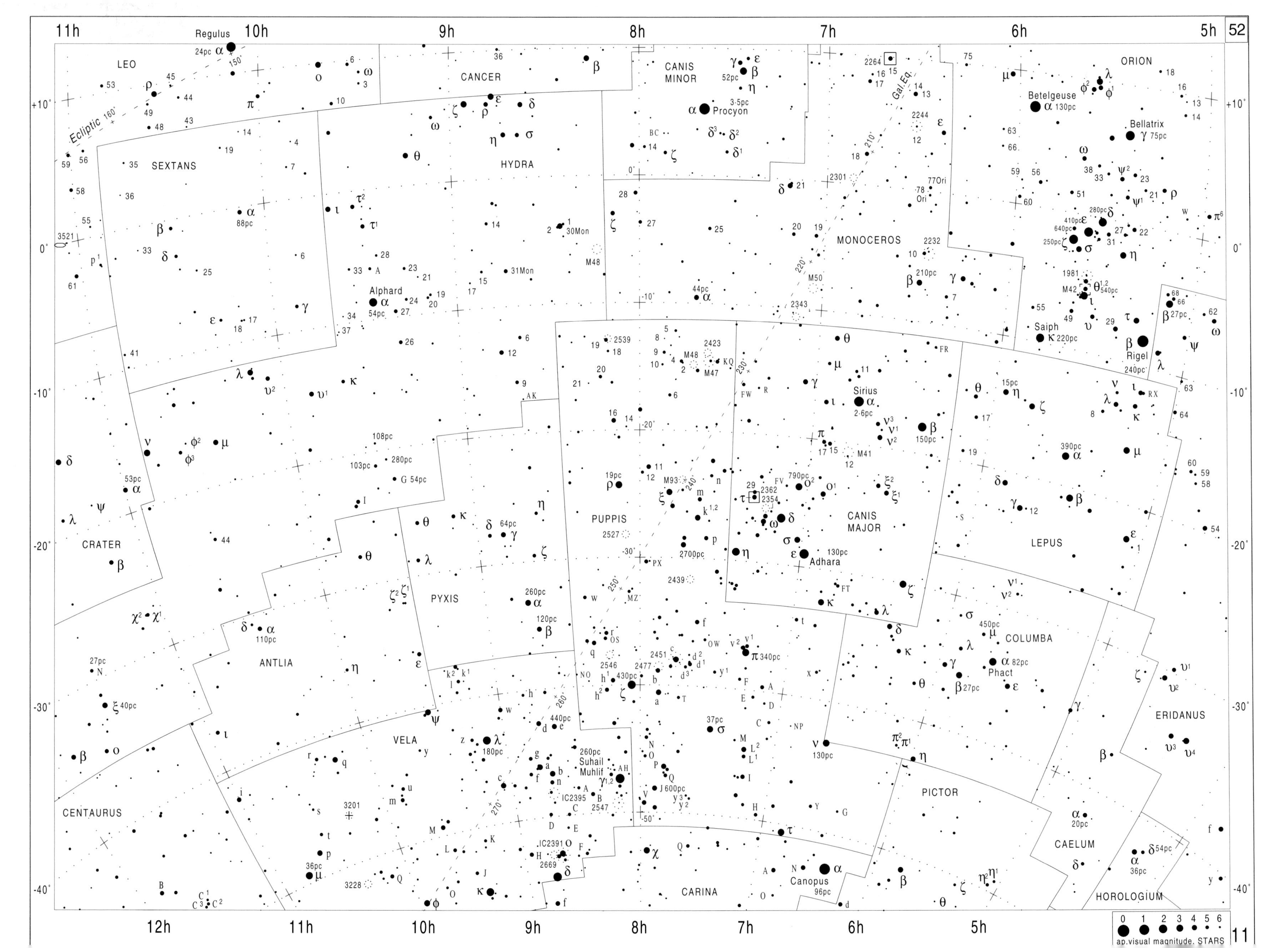

52
11h
10h
9h
8h
7h
6h
5h
LEO
CANCER
CANIS MINOR
ORION
SEXTANS
HYDRA
MONOCEROS
PUPPIS
CANIS MAJOR
LEPUS
CRATER
PYXIS
ANTLIA
COLUMBA
ERIDANUS
VELA
PICTOR
CAELUM
CENTAURUS
CARINA
HOROLOGIUM
Regulus
Procyon
Betelgeuse
Bellatrix
Alphard
Saiph
Rigel
Sirius
Adhara
Phact
Suhail
Muhlif
Canopus
Ecliptic
Gal.Eq.
+10°
0°
-10°
-20°
-30°
-40°
12h
0 1 2 3 4 5 6
ap.visual magnitude, STARS
11

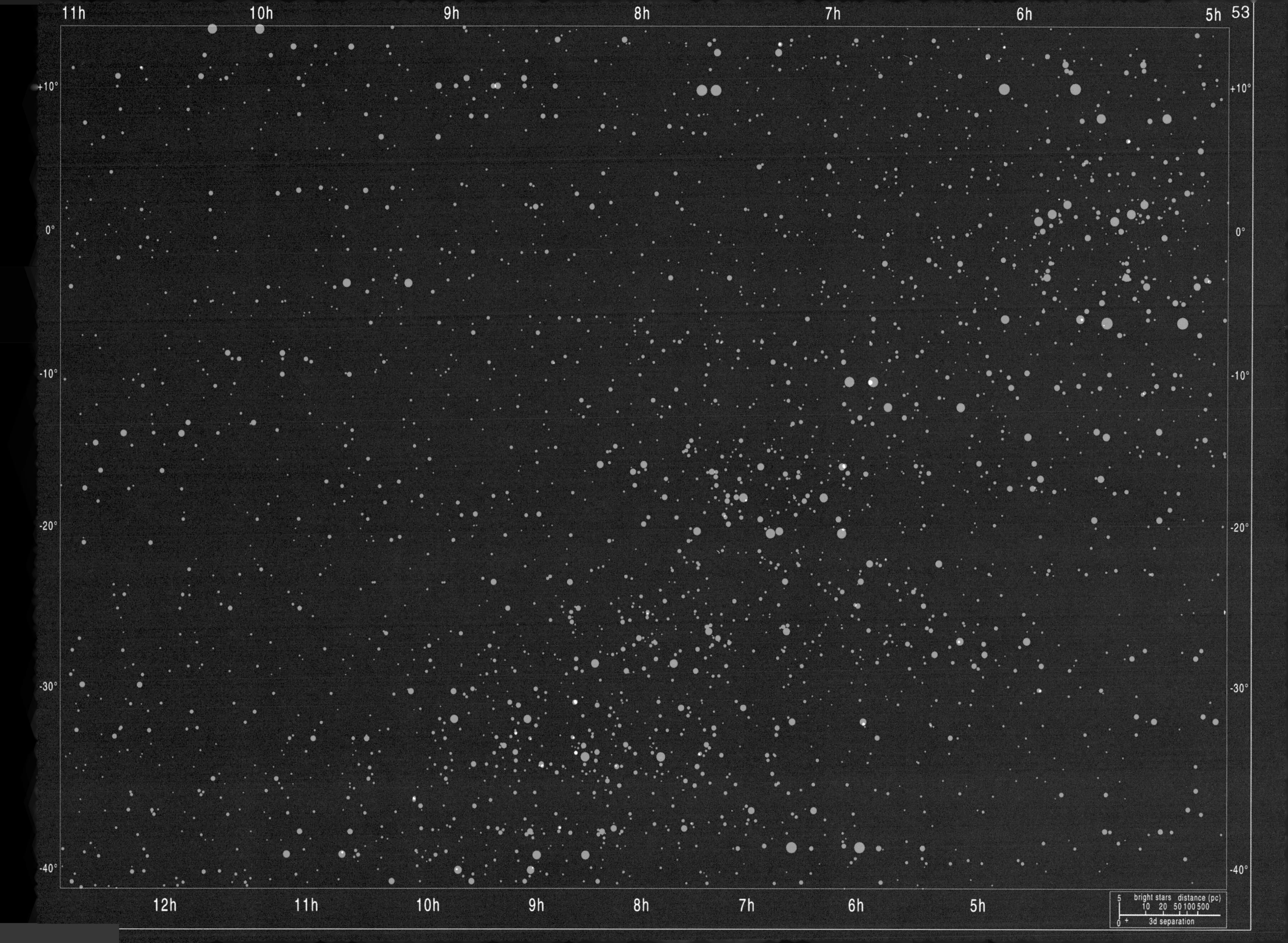
11h
10h
9h
8h
7h
6h
5h
+10°
0°
-10°
-20°
-30°
-40°
12h
11h
10h
9h
8h
7h
6h
5h
bright stars distance (pc)
5
10 20 50 100 500
0
3d separation

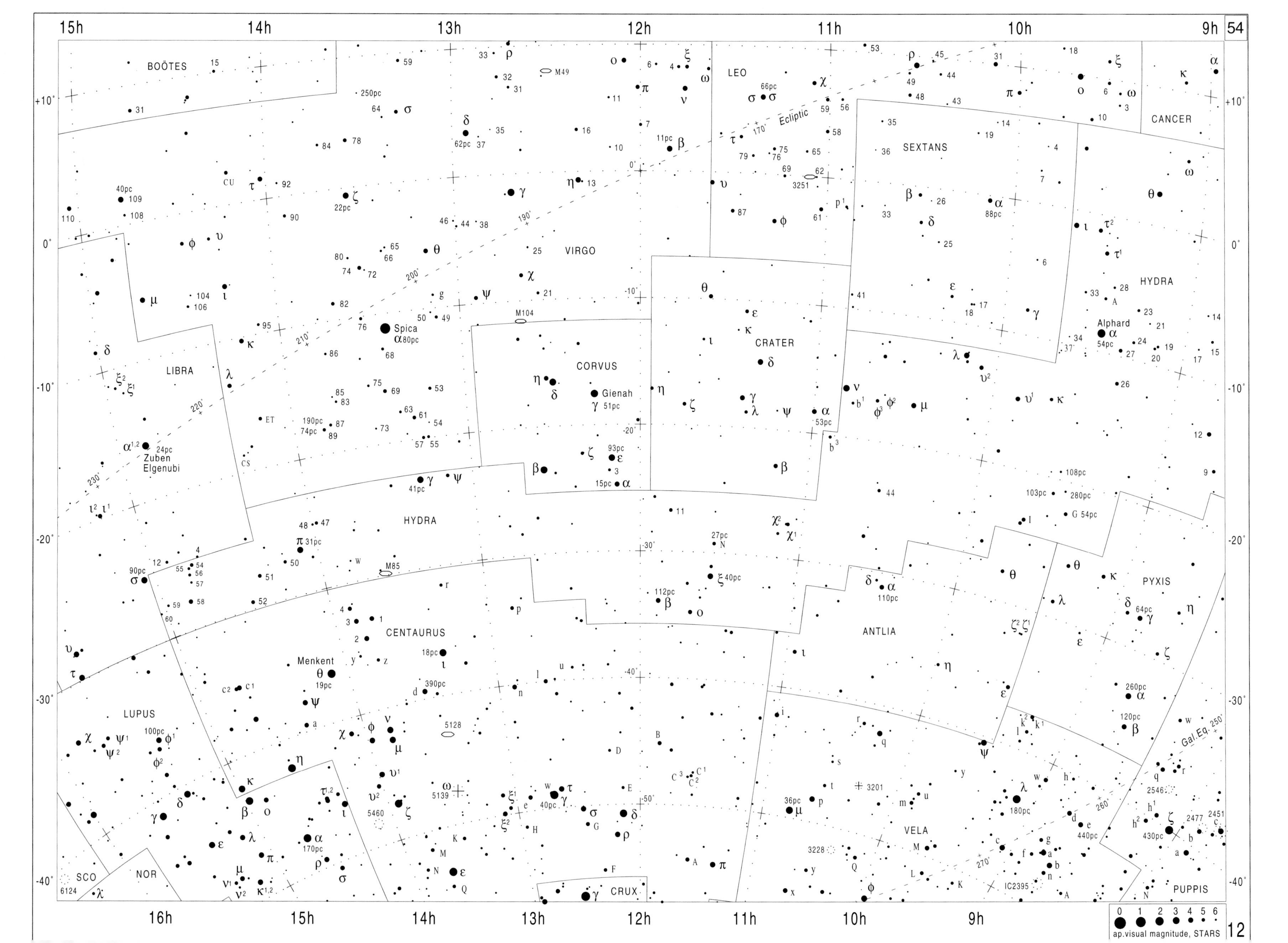
54
12
CANCER
HYDRA
PYXIS
PUPPIS
SEXTANS
ANTLIA
VELA
LEO
CRATER
VIRGO
CORVUS
CENTAURUS
CRUX
BOÖTES
LIBRA
LUPUS
NOR
SCO
Ecliptic
Gal.Eq.
Alphard
Spica
Gienah
Menkent
Zuben Elgenubi
M49
M104
M85
ap.visual magnitude, STARS

15h
14h
13h
12h
11h
10h
9h
+10°
0°
-10°
-20°
-30°
-40°
16h
15h
14h
13h
12h
11h
10h
9h
bright stars
distance (pc)
5
10 20 50 100 500
0
3d separation

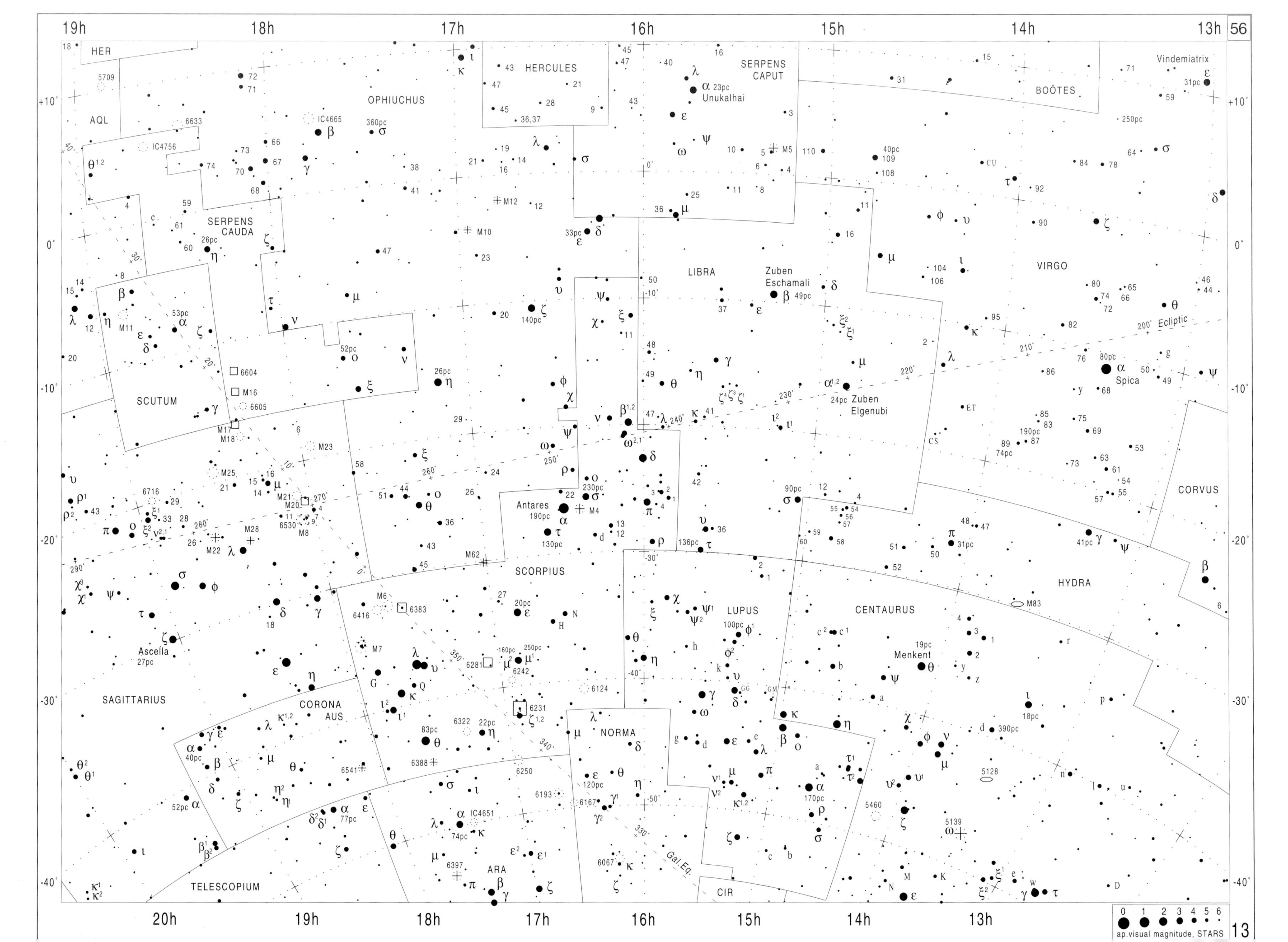
56
13
19h
18h
17h
16h
15h
14h
13h
20h
+10°
0°
-10°
-20°
-30°
-40°
HER
AQL
OPHIUCHUS
HERCULES
SERPENS CAPUT
Unukalhai
BOÖTES
Vindemiatrix
SERPENS CAUDA
LIBRA
Zuben Eschamali
VIRGO
Ecliptic
Spica
Zuben Elgenubi
SCUTUM
CORVUS
Antares
SCORPIUS
HYDRA
LUPUS
CENTAURUS
Menkent
M83
SAGITTARIUS
Ascella
CORONA AUS
NORMA
ARA
Gal.Eq.
CIR
TELESCOPIUM
0 1 2 3 4 5 6
ap.visual magnitude, STARS

19h
18h
17h
16h
15h
14h
13h
57
+10°
0°
-10°
-20°
-30°
-40°
20h
19h
18h
17h
16h
15h
14h
13h
5
bright stars
distance (pc)
10 20 50 100 500
3d separation

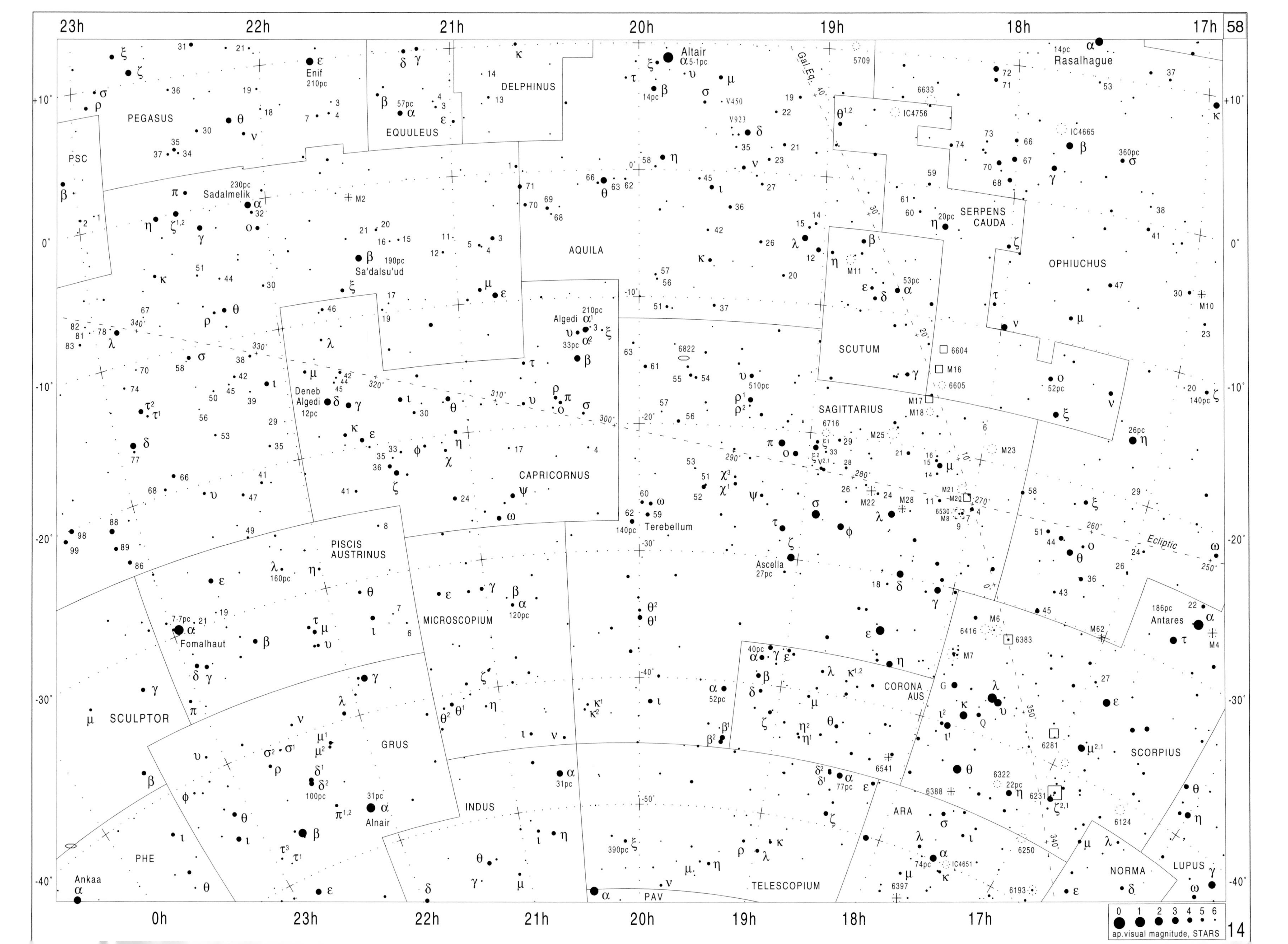

58
14
PEGASUS
PSC
EQUULEUS
DELPHINUS
AQUILA
SERPENS CAUDA
OPHIUCHUS
SCUTUM
SAGITTARIUS
CAPRICORNUS
PISCIS AUSTRINUS
MICROSCOPIUM
SCULPTOR
GRUS
INDUS
PHE
PAV
TELESCOPIUM
CORONA AUS
ARA
SCORPIUS
NORMA
LUPUS
Altair
Rasalhague
Enif
Sadalmelik
Sa'dalsu'ud
Algedi
Deneb Algedi
Terebellum
Ascella
Fomalhaut
Alnair
Ankaa
Antares
Ecliptic
Gal.Eq.
ap.visual magnitude, STARS

+10°
0°
-10°
-20°
-30°
-40°
0h
23h
22h
21h
20h
19h
18h
17h
bright stars distance (pc)
5
10 20 50 100 500
0
3d separation

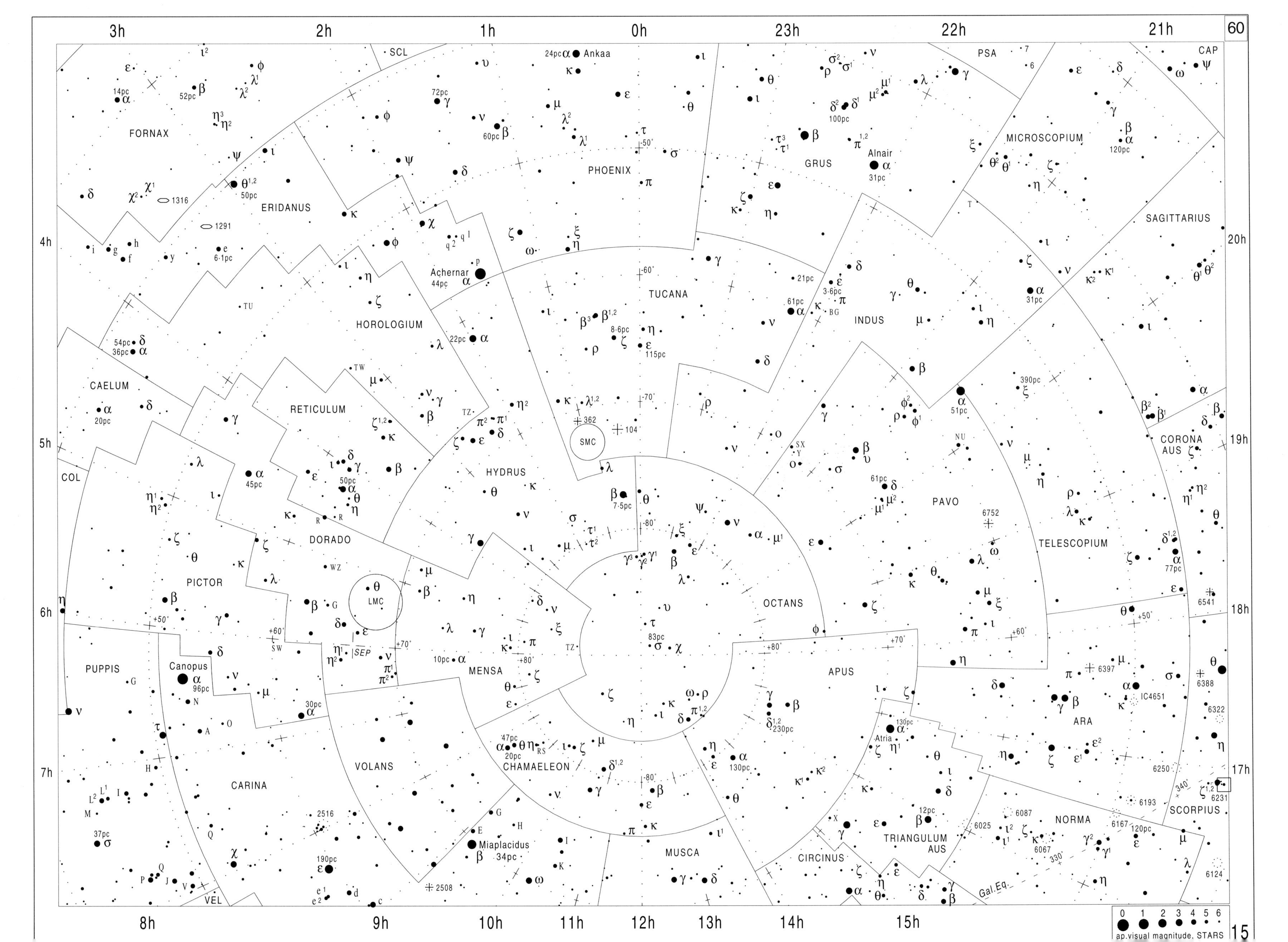

3h
2h
1h
0h
23h
22h
21h
20h
19h
18h
17h
4h
5h
6h
7h
8h
9h
10h
11h
12h
13h
14h
15h
FORNAX
ERIDANUS
SCL
PHOENIX
GRUS
PSA
MICROSCOPIUM
CAP
SAGITTARIUS
CAELUM
HOROLOGIUM
TUCANA
INDUS
RETICULUM
HYDRUS
PAVO
TELESCOPIUM
CORONA AUS
COL
DORADO
PICTOR
OCTANS
MENSA
PUPPIS
CARINA
VOLANS
CHAMAELEON
APUS
ARA
SCORPIUS
NORMA
TRIANGULUM AUS
CIRCINUS
MUSCA
VEL
Ankaa
Achernar
Alnair
Canopus
Miaplacidus
Atria
SMC
LMC
SEP
Gal.Eq.
ap.visual magnitude, STARS
0 1 2 3 4 5 6

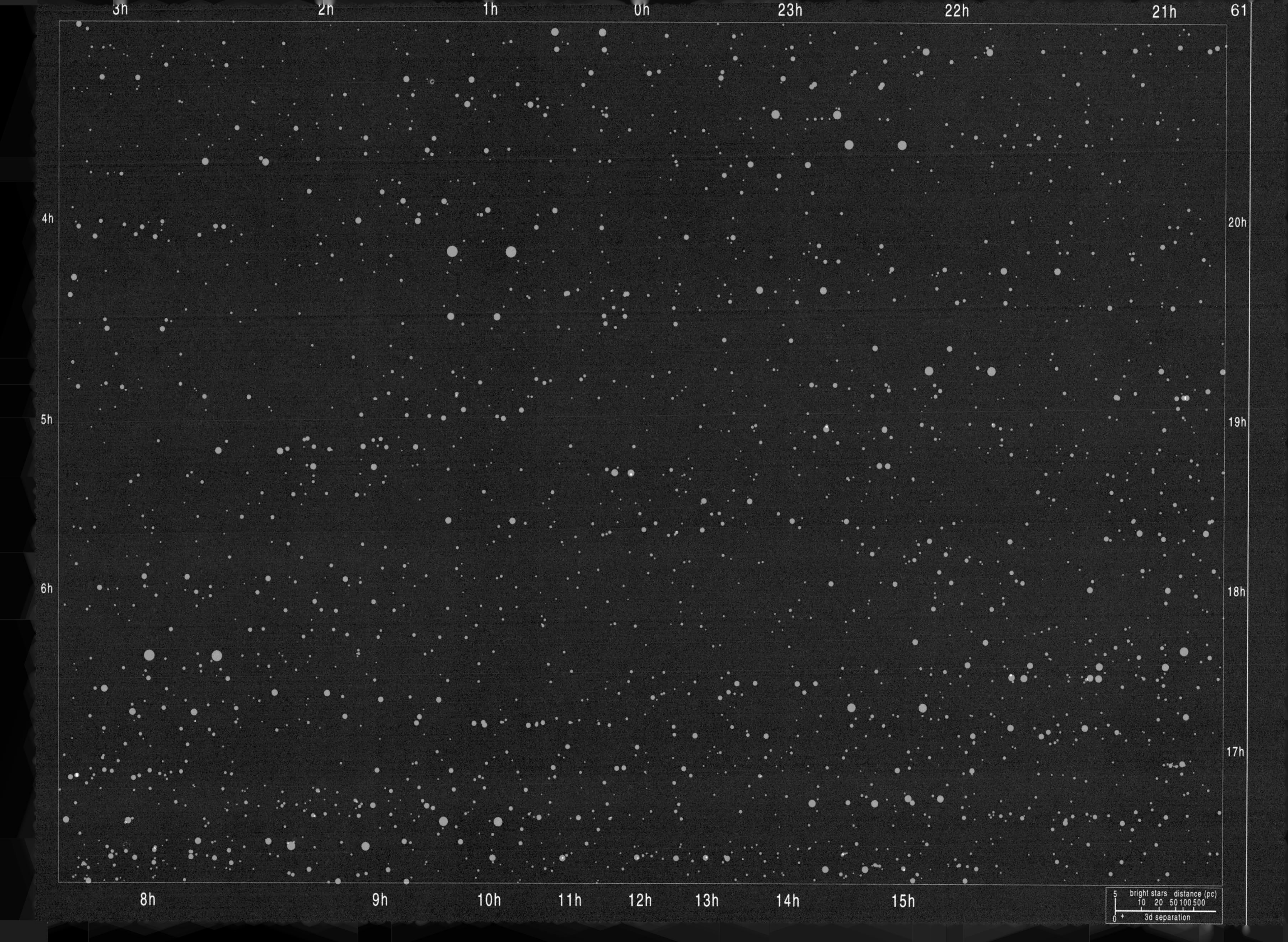
3h
2h
1h
0h
23h
22h
21h
4h
5h
6h
20h
19h
18h
17h
8h
9h
10h
11h
12h
13h
14h
15h
bright stars
distance (pc)
5
10
20
50
100
500
0
3d separation

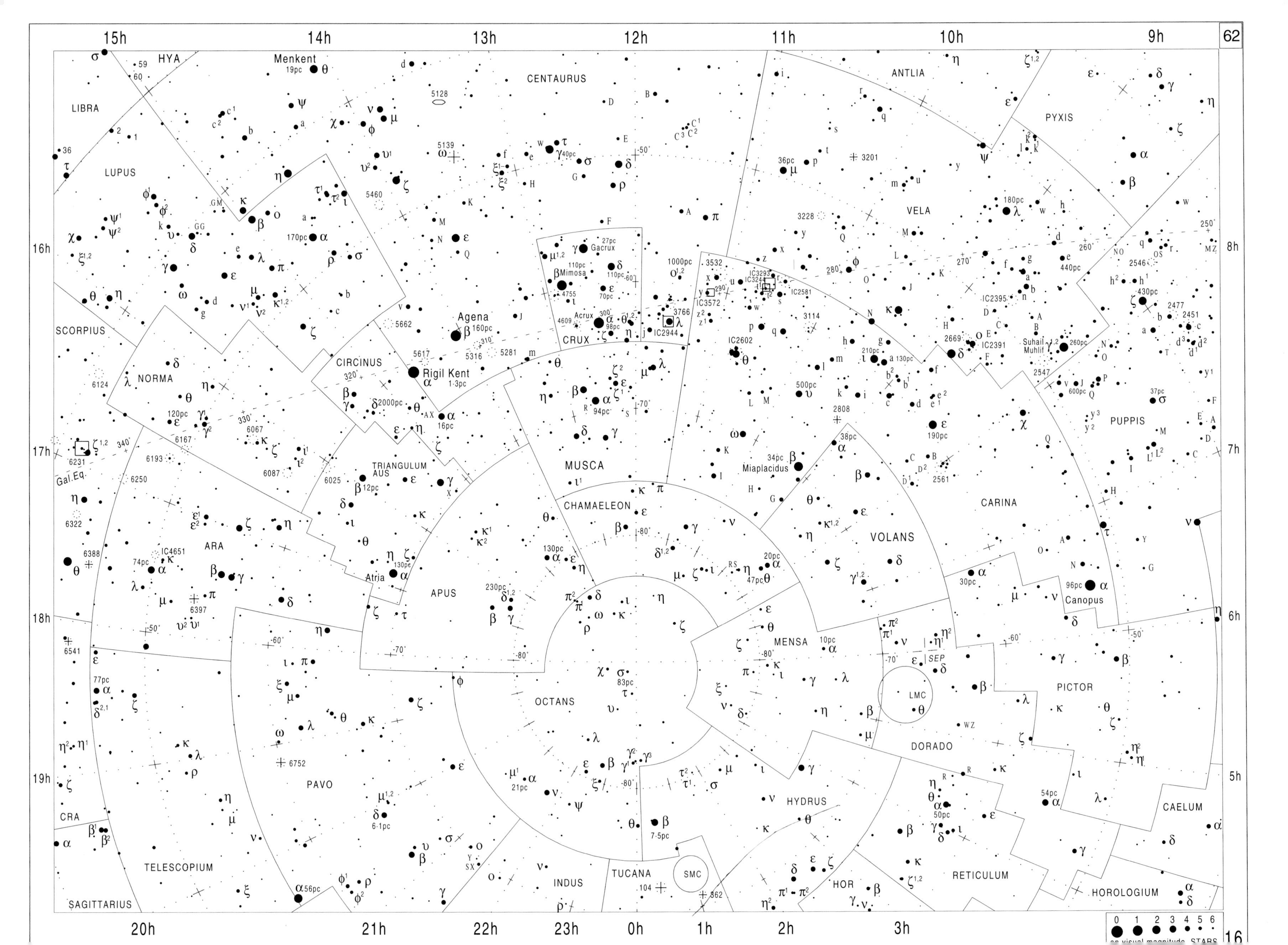
HYA
LIBRA
LUPUS
SCORPIUS
NORMA
CIRCINUS
CENTAURUS
CRUX
MUSCA
CHAMAELEON
ANTLIA
PYXIS
VELA
PUPPIS
CARINA
VOLANS
MENSA
PICTOR
CAELUM
DORADO
HYDRUS
RETICULUM
HOROLOGIUM
HOR
TUCANA
INDUS
OCTANS
APUS
TRIANGULUM AUS
ARA
PAVO
TELESCOPIUM
CRA
SAGITTARIUS
Menkent
Agena
Rigil Kent
Gacrux
Mimosa
Acrux
Atria
Miaplacidus
Canopus
Suhail
Muhlif
LMC
SMC
SEP
Gal.Eq.
STARS

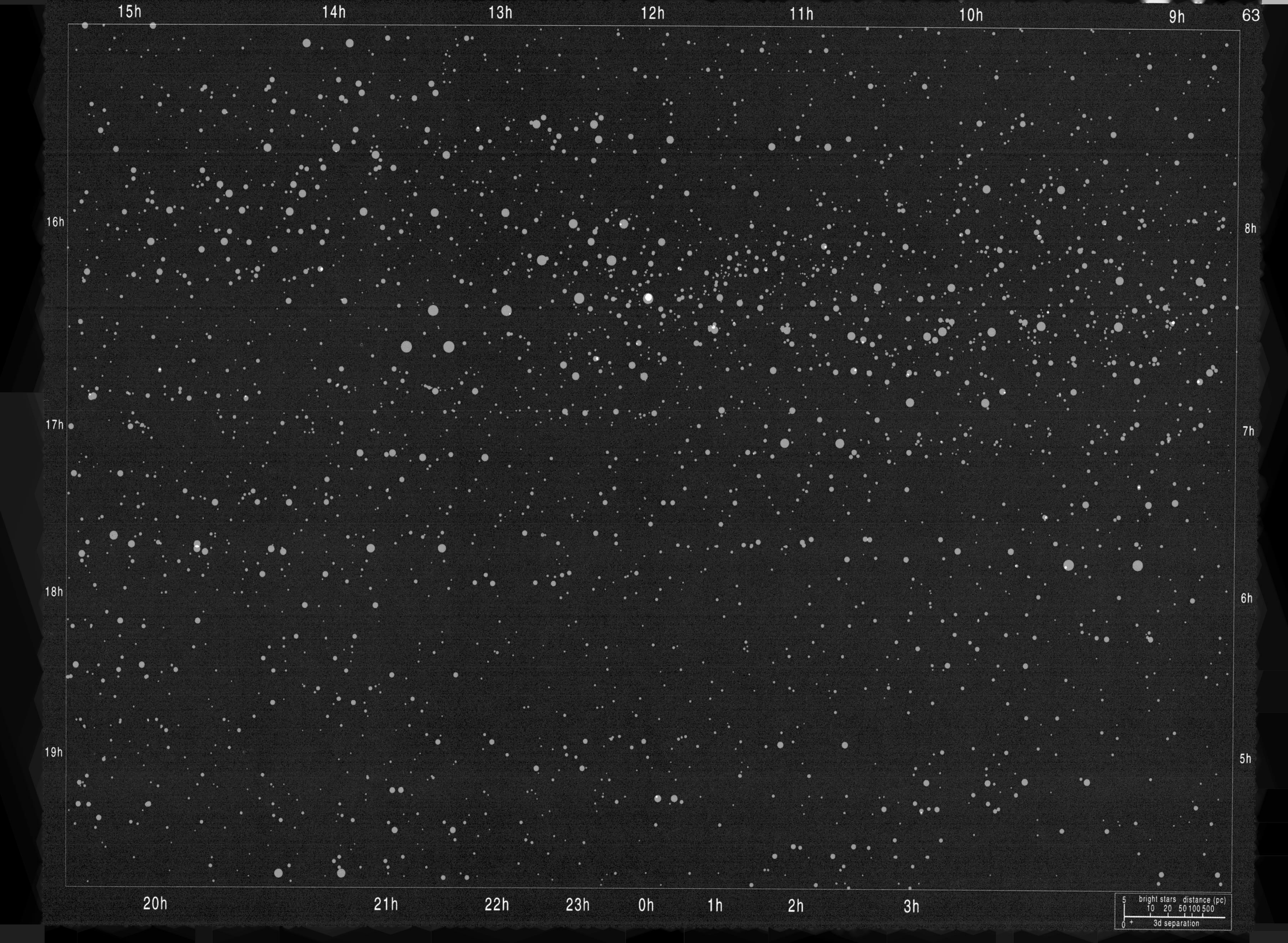
15h
14h
13h
12h
11h
10h
9h
16h
17h
18h
19h
8h
7h
6h
5h
20h
21h
22h
23h
0h
1h
2h
3h
bright stars distance (pc)
5
10 20 50 100 500
0
3d separation

# Introduction

# The Galaxy Maps

The Galaxy Maps are plotted from the 1998 September 16th version of *the CfA Redshift Survey and Catalogue of Galaxies*. The key to the maps assumes a Hubble constant $H_0 = 65{\cdot}202$ km per second per Mpc, such that each kilometre per second of recession velocity equates to a separation of 15,337 pc.

Outside of a general recession all galaxies have their own individual motions relative to other galaxies. A small number of galaxies close to the Milky Way show negative velocities—they are getting nearer rather than farther away. Because objects with negative velocity cannot be plotted on maps where distance is related to redshift, these galaxies have been excluded from the maps (examples are M98 and M86). These exclusions highlight the fact that redshift is not a reliable indicator of distance for the nearest galaxies. The Large and Small Magellanic Clouds (*LMC, SMC)* are not shown on the stereographs, but they are shown on the annotated maps to give a location.

The Galaxy Maps are shown in a projection centred on the so-called 'galactic poles' (elsewhere shown as *NGP* and *SGP* on Bright Star Maps 6 and 9). The galactic poles are found at 90° to the *Galactic Equator*, the plane of the Milky Way Galaxy itself. The Sun and Solar System is located within the plane of the Milky Way Galaxy, about 10 kpc out from the galactic centre observed in the direction of *Sagittarius*, galactic longitude 0°. Most visible stars and obscuring clouds of gas and dust local to our Galaxy are found in the galactic plane. They shut out the light of more distant galaxies to form a 'plane of obscuration' 40° to 60° deep, seen along the foot of maps 2–5 and the head of maps 6–9. The seeming rarity of galaxies in this region is an illusion, an artifact created by the conditions of observation.

The *CfA Redshift Catalogue* is compiled from more than fifty sources. Individual observatories and observation programmes observe galaxies in particular areas of the sky to different limits and criteria of measurement and recording, so the completeness of the catalogue varies.

After removal of disallowed sources, and with the exception of objects listed from four other sources, the maps plot all objects in the 1998 version of the catalogue to a magnitude limit of 15·99. The galaxy fields from the four source catalogues numbered 27, 36, 40, and 51 all appear to have been surveyed to an untypical level of completeness. The boundaries of the survey fields matched declination bands on trial plots, and risked producing observation artifacts that could be mistaken for the contours of natural clustering. For this reason, objects listed from these source catalogues have been restricted to magnitude 14·99. Other congregations of object plots seen on the maps might represent observation artifacts, might describe real clusters, or might be some combination of the two. Right ascension and declination coordinates are given for the approximate centre of some of these plot groupings.

A feature of the annotated maps is the conventional constellation boundaries, which give a sense of the more familiar celestial coordinate system, and allow the major clusters of galaxies to be located against the regions of the sky after which they are frequently named. However, the dedicated naming of individual galaxies and clusters of galaxies is restricted to a few well-established examples. A limited annotation concentrates on brighter and nearer objects. Numbers without prefix are NGC numbers; prefix 'IC' denotes Index Catalogue numbers, prefix 'M' denotes Messier numbers, and prefix 'R' *CfA Redshift Catalogue* numbers. The compression used in the representation of depth follows a logarithmic curve. The page distance is set at 10 Mpc.

## Some Interesting Structures

The most fascinating thing about illustrating galaxies in 3-D is the way it shows up elements of structure in the way that the galaxies are distributed. Such structure is particularly easy to see among clusters of fainter and (by implication) more distant objects. In preparing the maps it was possible to show objects to a fair degree of completeness, perhaps magnitude 14, and lose some of the structure, or else to show as much structure as possible and lose completeness. We chose the second option.

The dominant feature of the northern galactic region is the line of galaxies running from *Ursa Major* to *Centaurus* (seen running diagonally across the North Circle map), and concentrated in the region running through *Virgo* (map 1). This line of galaxies has been given various names, including 'The Great Wall' and 'The Supergalactic Equator'. The concentration of galaxies along this axis is undeniable, although the limit of representation at one-tenth the speed of light means that this area represents what is still a comparatively small volume of space in the universe.

The major structure in the southern region is equally obvious. A line of galaxies runs in an arc from *Perseus* (map 8) through *Andromeda* and *Pegasus* to *Lacerta* (map 7). Another filament runs from *Cetus* (map 7) through *Pisces*, with a nearer cluster between the two filaments in *Pegasus*. How many filaments are there? Richard Monkhouse sees three strands along the northern side and favours the name 'Twisted Rope' to describe this particular element. John Cox favours the name 'Southern Wall' to cover the whole of the larger grouping. There are elements of bridging structures running through *Aires* and (celestial) east *Pisces* (map 8); in some ways the overall structure resembles a short piece of model DNA. The south side of the structure has a clean finish, a chasm, another filament, then a void with concentrated clusters in *Phoenix*.

Beside these major structures, the maps abound with smaller-scale groupings and filaments. One such filament appears to run from behind the *Coma Berenices* cluster down to a cluster in *Corona Borealis*. There may or may not be a string running across to a cluster in *Serpens*.

When viewing the stereographs, you should bear in mind that the depth scale is extremely compressed. What can look like layers or shells in the perspective of the stereographs could in reality be radial structures, streaming away from the observer's viewpoint to a much greater extent than wrapping around. There is no getting around this flattening effect, and the reader has to develop a sense of logarithmic depth to allow for it.

The observer's viewpoint in the stereographs is the Milky Way Galaxy, whose obscuring effect has already been mentioned. What the Milky Way hides can only be guessed at. The Milky Way is itself a member of a so-called 'Local Group' that is lost in the present projection, and that group is in its turn an outlying member of the *Virgo Cluster* (map 1). By viewing this area in a relaxed way the reader will find a small fraction breaks away and floats upwards, these galaxies forming our connection to the larger grouping beyond.

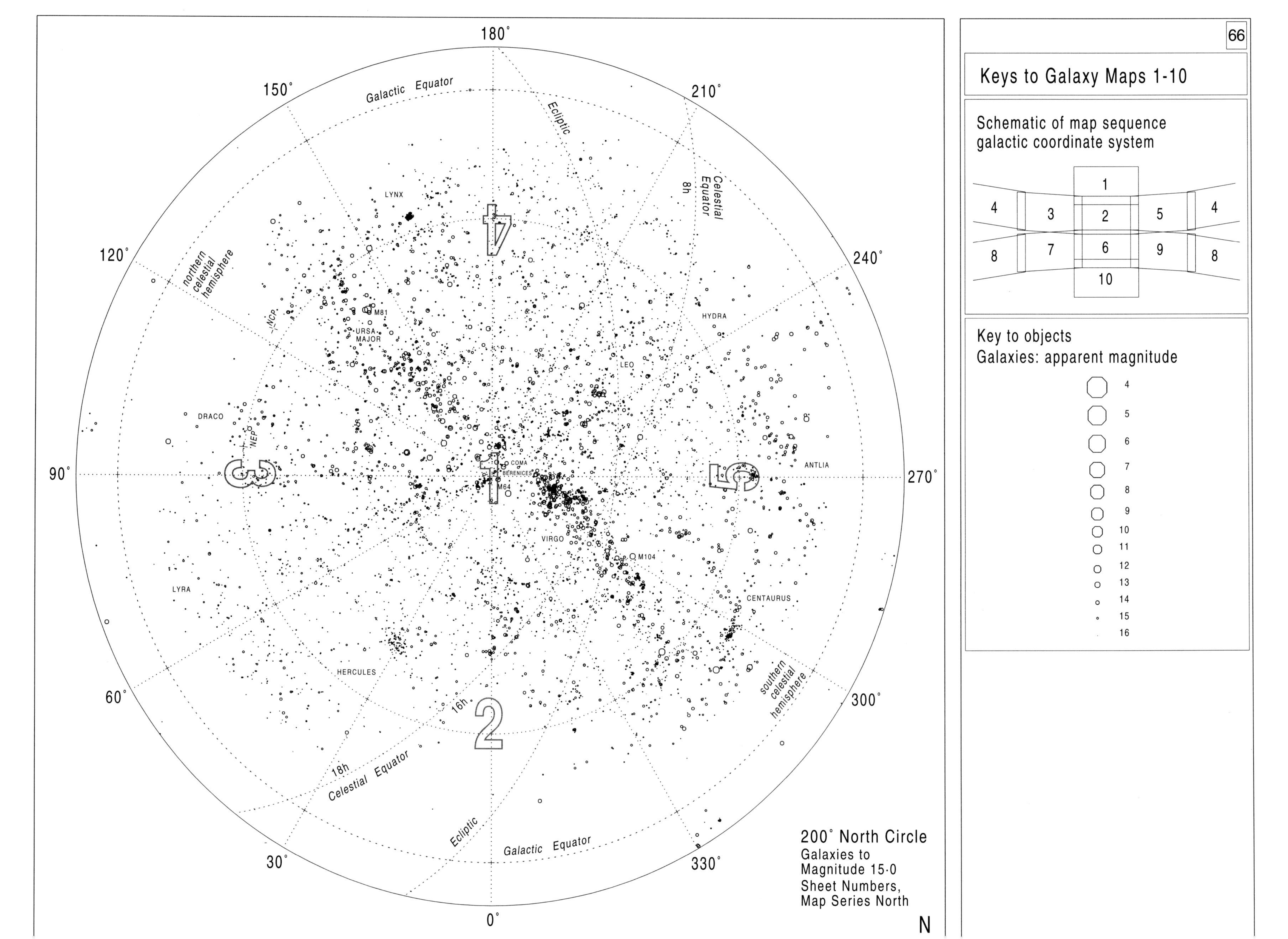


## Keys to Galaxy Maps 1-10

### Schematic of map sequence galactic coordinate system



### Key to objects

Galaxies: apparent magnitude

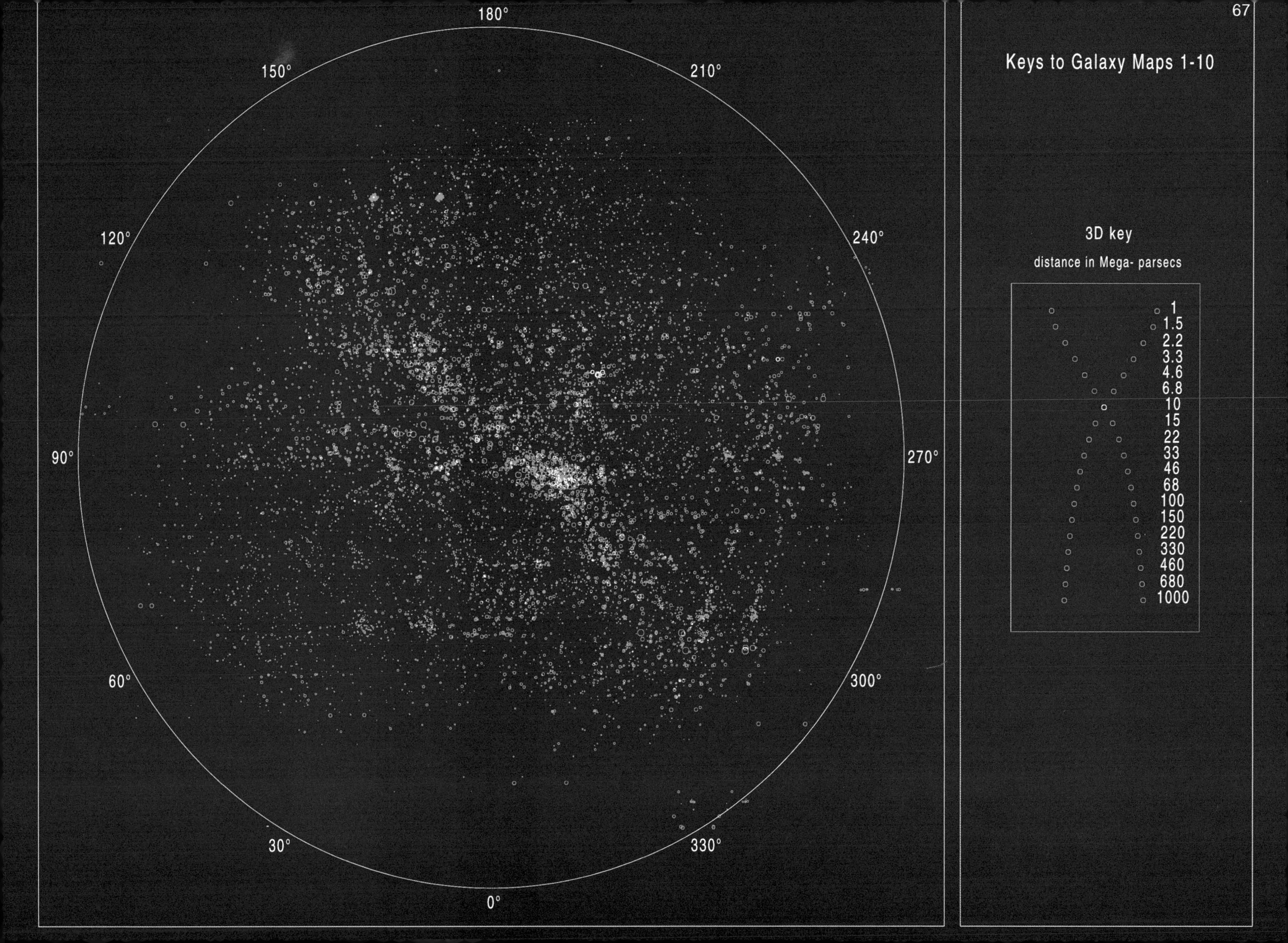
180°
150°
210°
120°
240°
90°
270°
60°
300°
30°
330°
0°
Keys to Galaxy Maps 1-10
3D key
distance in Mega- parsecs
1
1.5
2.2
3.3
4.6
6.8
10
15
22
33
46
68
100
150
220
330
460
680
1000

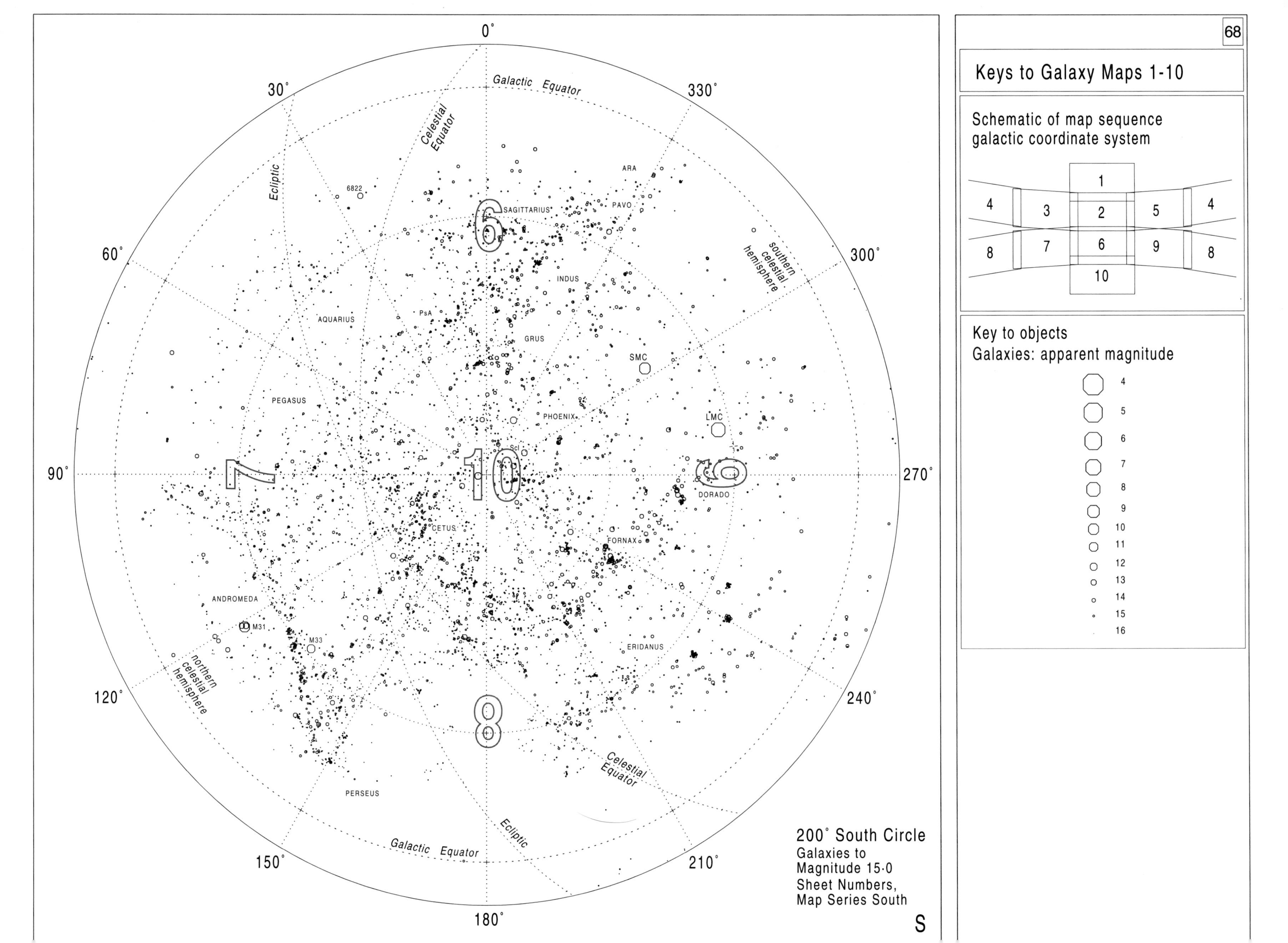


## Keys to Galaxy Maps 1-10

Schematic of map sequence
galactic coordinate system



Key to objects
Galaxies: apparent magnitude

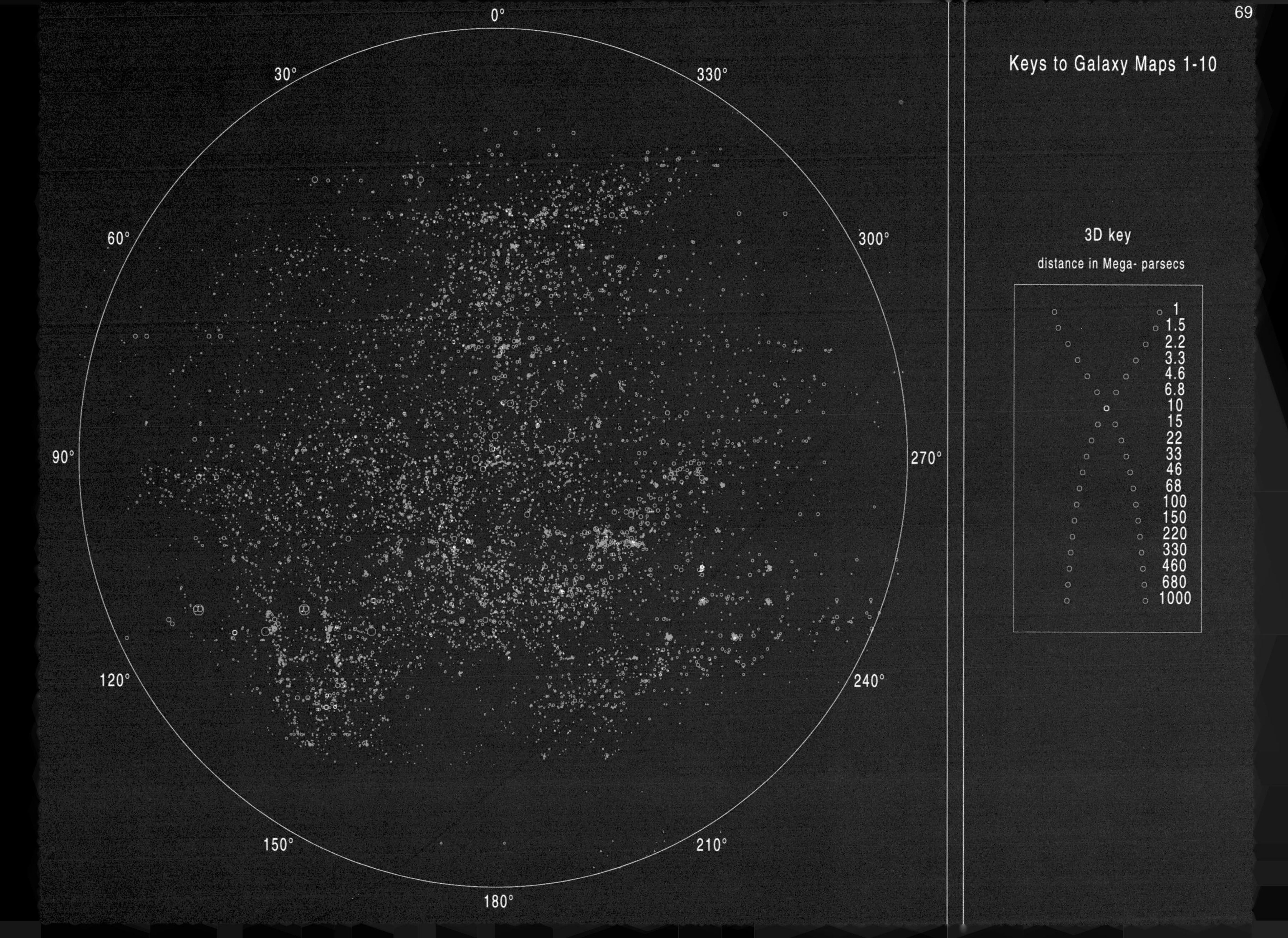
0°
30°
330°
60°
300°
90°
270°
120°
240°
150°
210°
180°
Keys to Galaxy Maps 1-10
3D key
distance in Mega- parsecs
1
1.5
2.2
3.3
4.6
6.8
10
15
22
33
46
68
100
150
220
330
460
680
1000

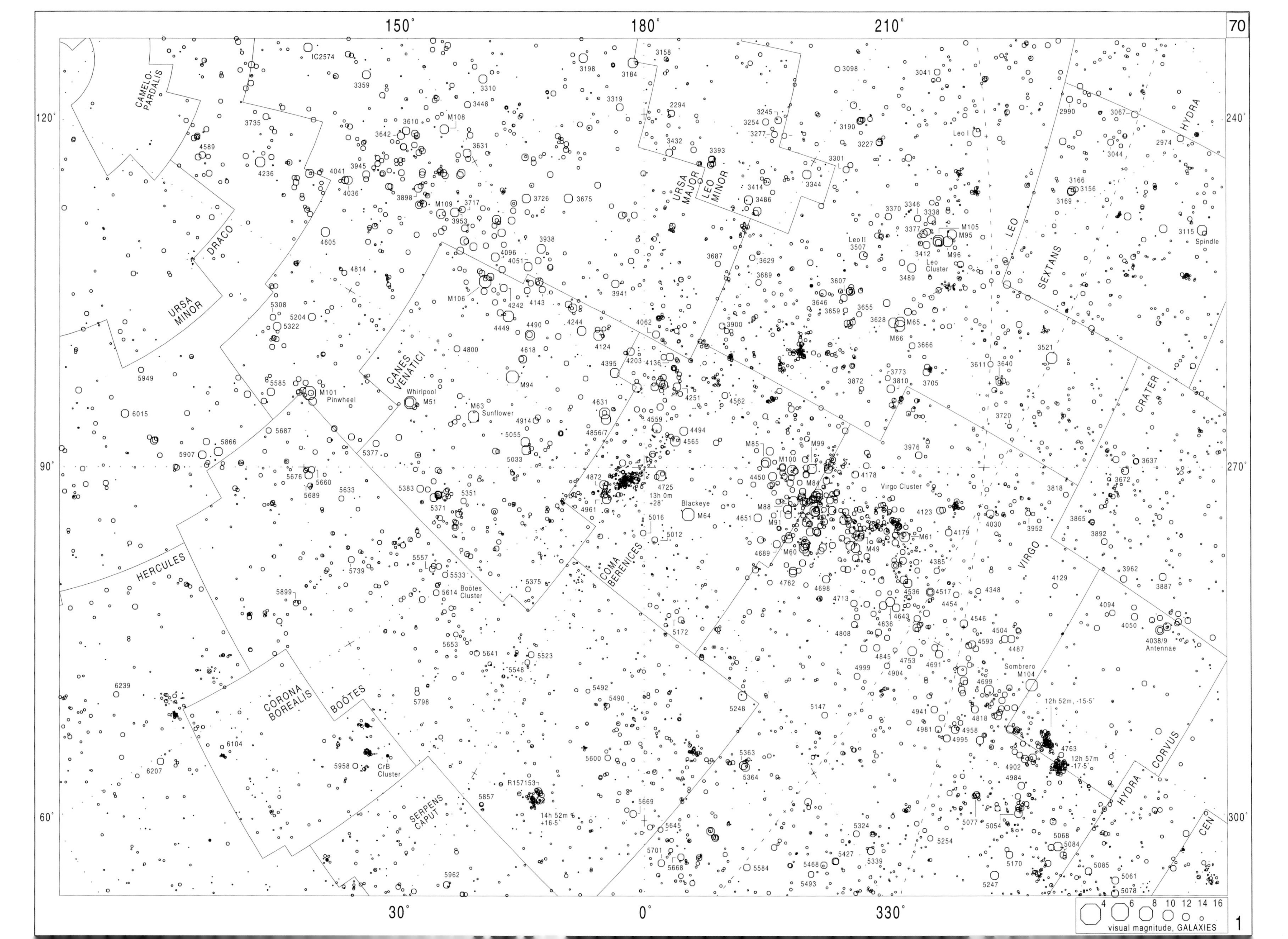
150°
180°
210°
120°
240°
90°
270°
60°
300°
30°
0°
330°
CAMELOPARDALIS
DRACO
URSA MINOR
HERCULES
CORONA BOREALIS
BOÖTES
SERPENS CAPUT
CANES VENATICI
COMA BERENICES
URSA MAJOR
LEO MINOR
LEO
SEXTANS
HYDRA
CRATER
VIRGO
CORVUS
CEN
Spindle
Leo I
Leo II
Leo Cluster
M105
M95
M96
M65
M66
M108
M109
M106
M94
M63 Sunflower
M51 Whirlpool
M101 Pinwheel
M64 Blackeye
13h 0m +28°
M85
M99
M100
M84
M88
M91
M60
M49
M61
Virgo Cluster
Boötes Cluster
CrB Cluster
14h 52m +16.5°
R157153
Sombrero M104
12h 52m, -15.5°
12h 57m -17.5°
4038/9 Antennae
4 6 8 10 12 14 16
visual magnitude, GALAXIES

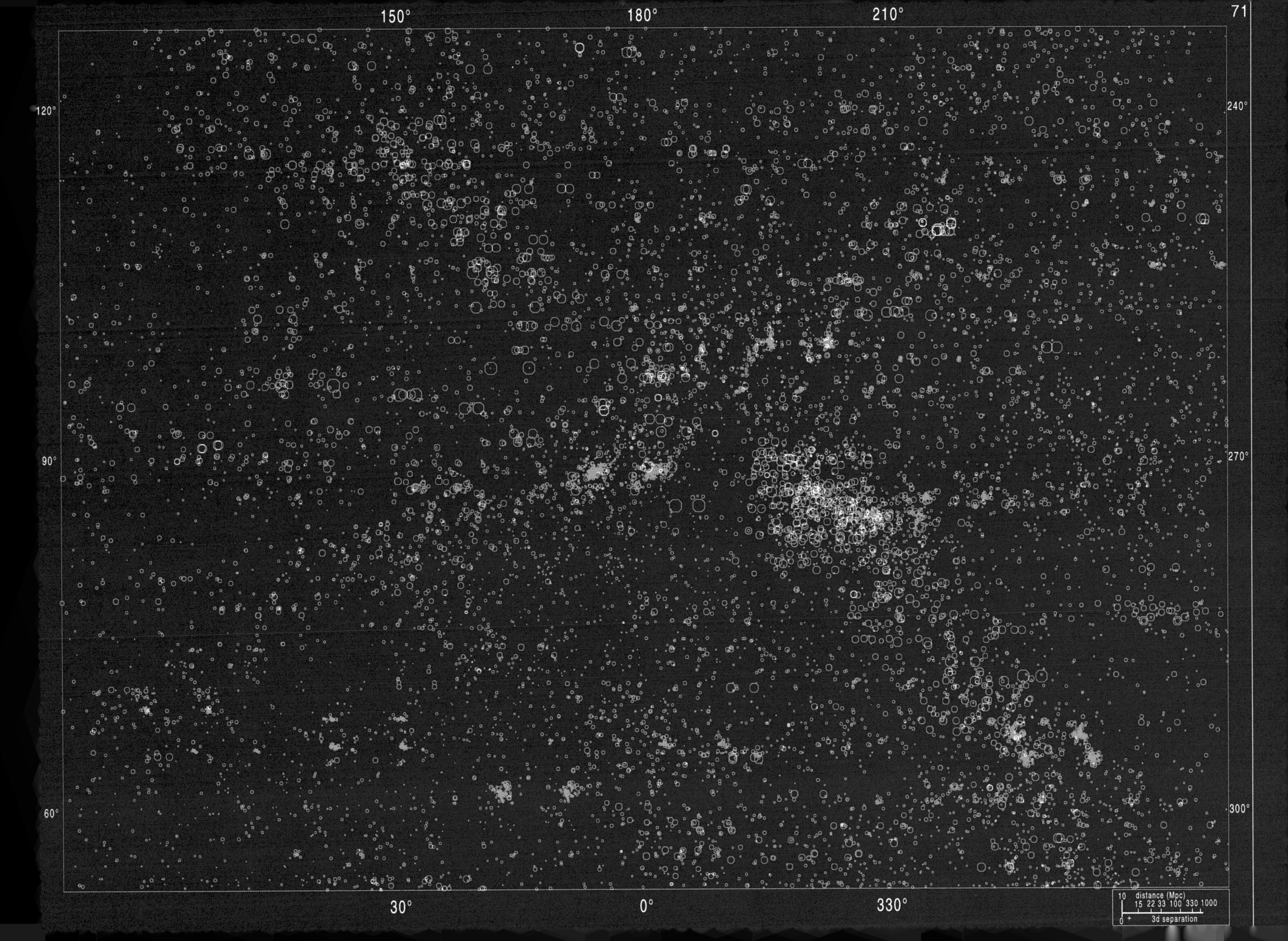

150°
180°
210°
120°
240°
90°
270°
60°
300°
30°
0°
330°
distance (Mpc)
10 15 22 33 100 330 1000
0 3d separation

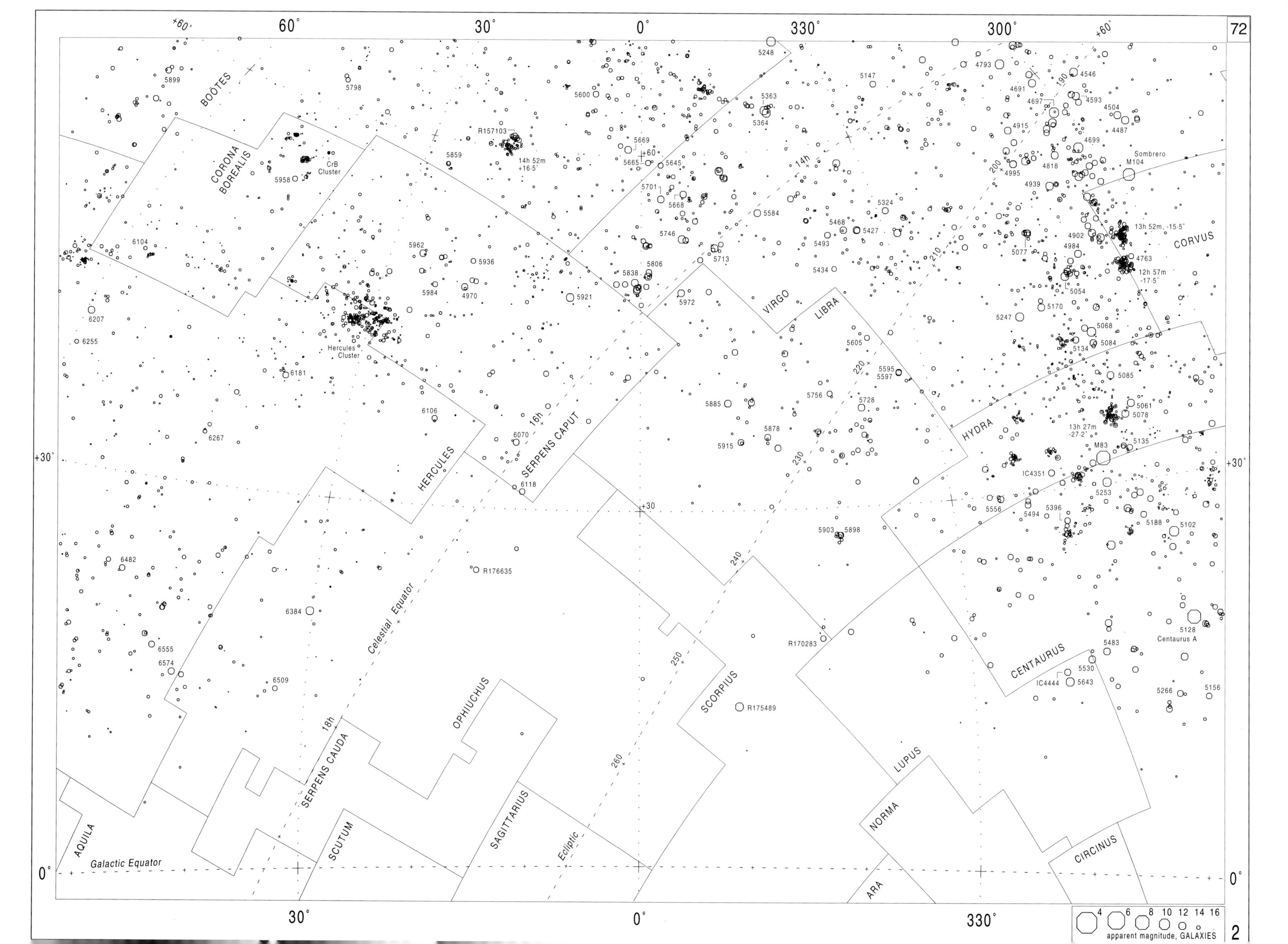

BOOTES
CORONA BOREALIS
CrB Cluster
HERCULES
Hercules Cluster
SERPENS CAPUT
VIRGO
LIBRA
CORVUS
HYDRA
CENTAURUS
Centaurus A
Sombrero
M104
M83
SCORPIUS
OPHIUCHUS
SERPENS CAUDA
SCUTUM
SAGITTARIUS
AQUILA
LUPUS
NORMA
ARA
CIRCINUS
Celestial Equator
Ecliptic
Galactic Equator
14h 52m +16.5°
13h 52m, -15.5°
12h 57m -17.5°
13h 27m -27.2°
14h
16h
18h
R157103
R176635
R170283
R175489
IC4351
IC4444
60°
30°
0°
330°
300°
+60°
+30°
apparent magnitude, GALAXIES
4 6 8 10 12 14 16

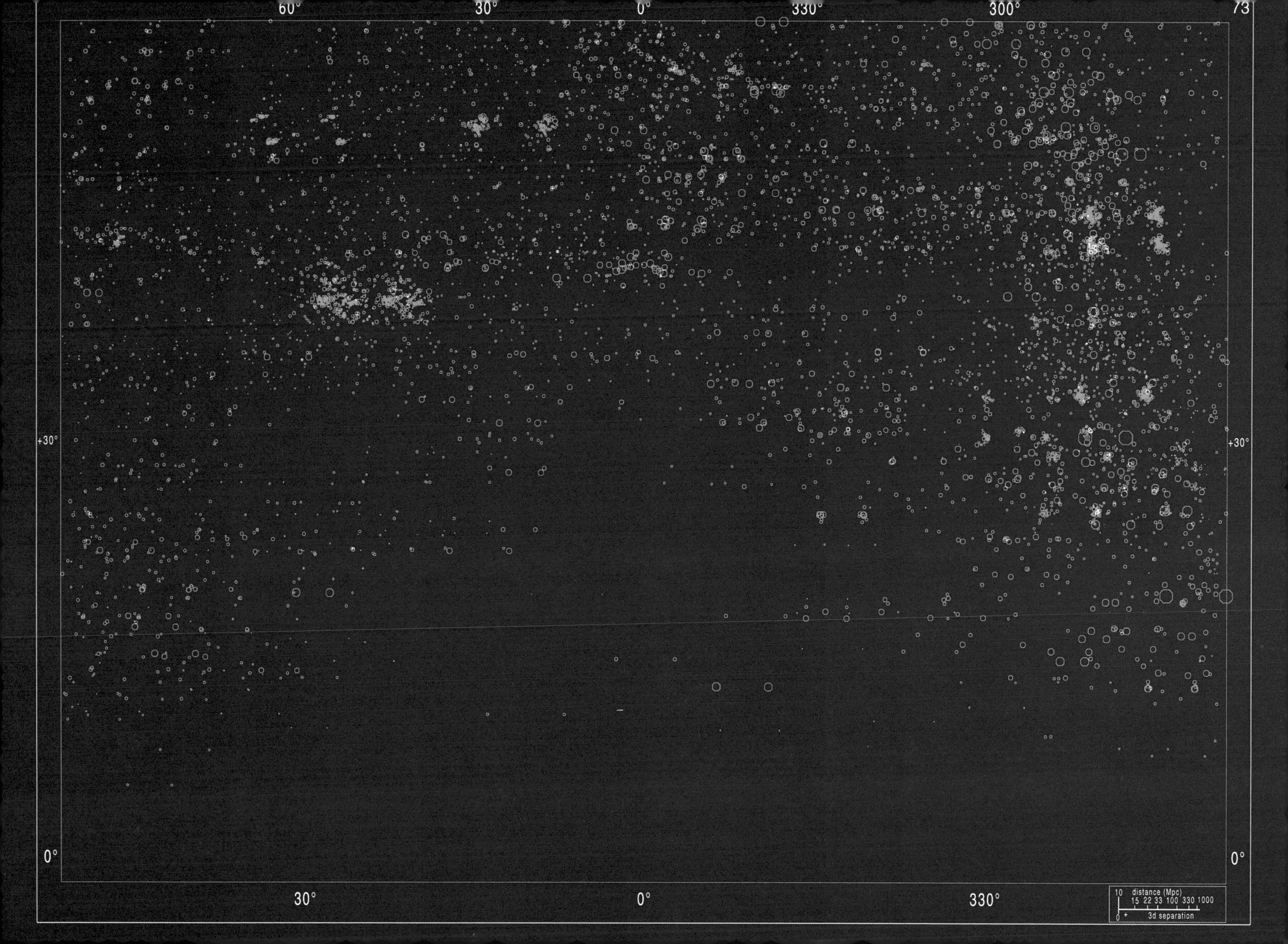
60°
30°
0°
330°
300°
+30°
+30°
0°
0°
30°
0°
330°
distance (Mpc)
10 15 22 33 100 330 1000
0
3d separation

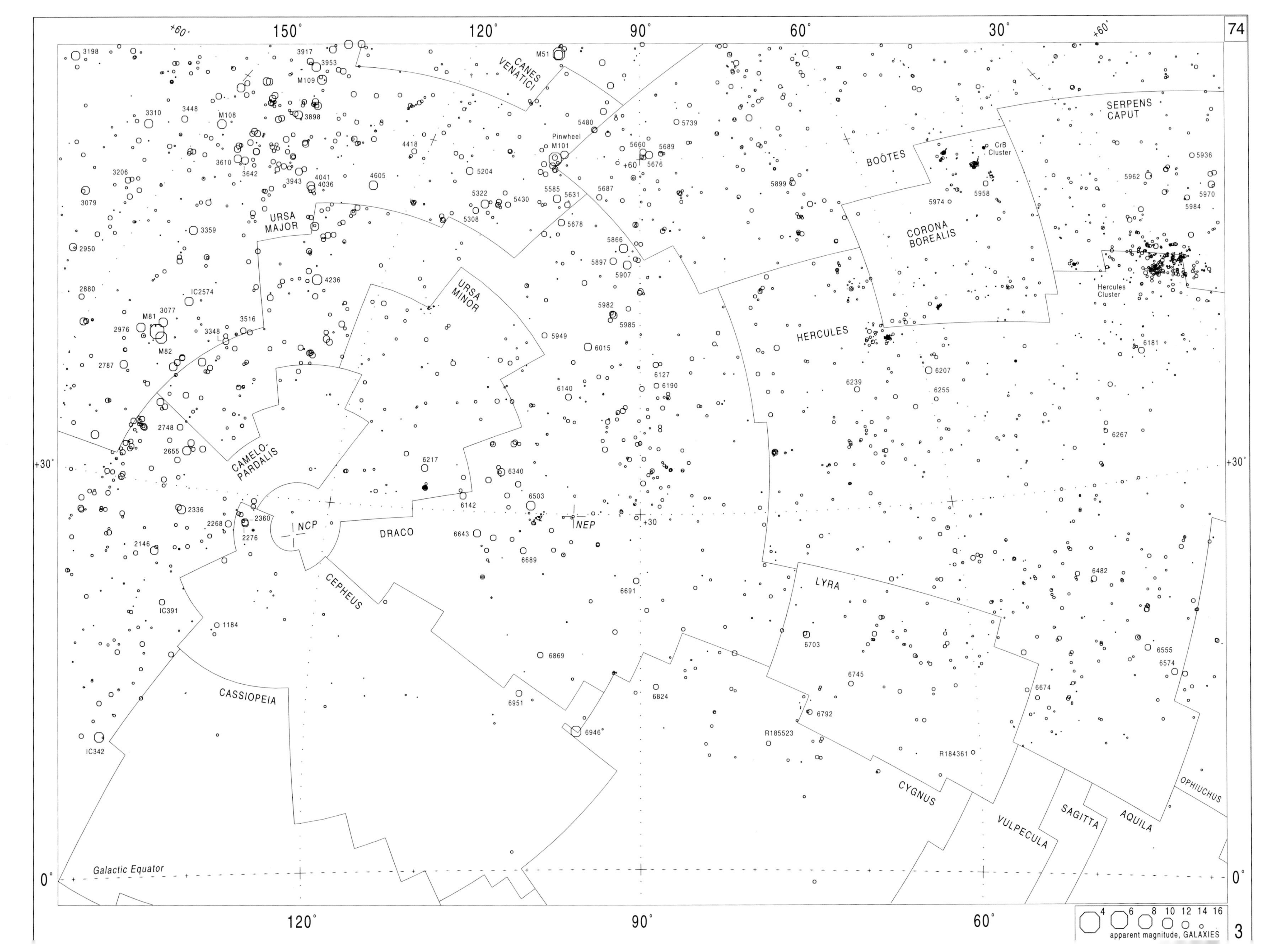

SERPENS CAPUT
BOÖTES
CORONA BOREALIS
CrB Cluster
HERCULES
Hercules Cluster
CANES VENATICI
URSA MAJOR
URSA MINOR
CAMELOPARDALIS
DRACO
CEPHEUS
CASSIOPEIA
LYRA
CYGNUS
VULPECULA
SAGITTA
AQUILA
OPHIUCHUS
NCP
NEP
Pinwheel M101
M51
M109
M108
M81
M82
Galactic Equator
apparent magnitude, GALAXIES
4 6 8 10 12 14 16

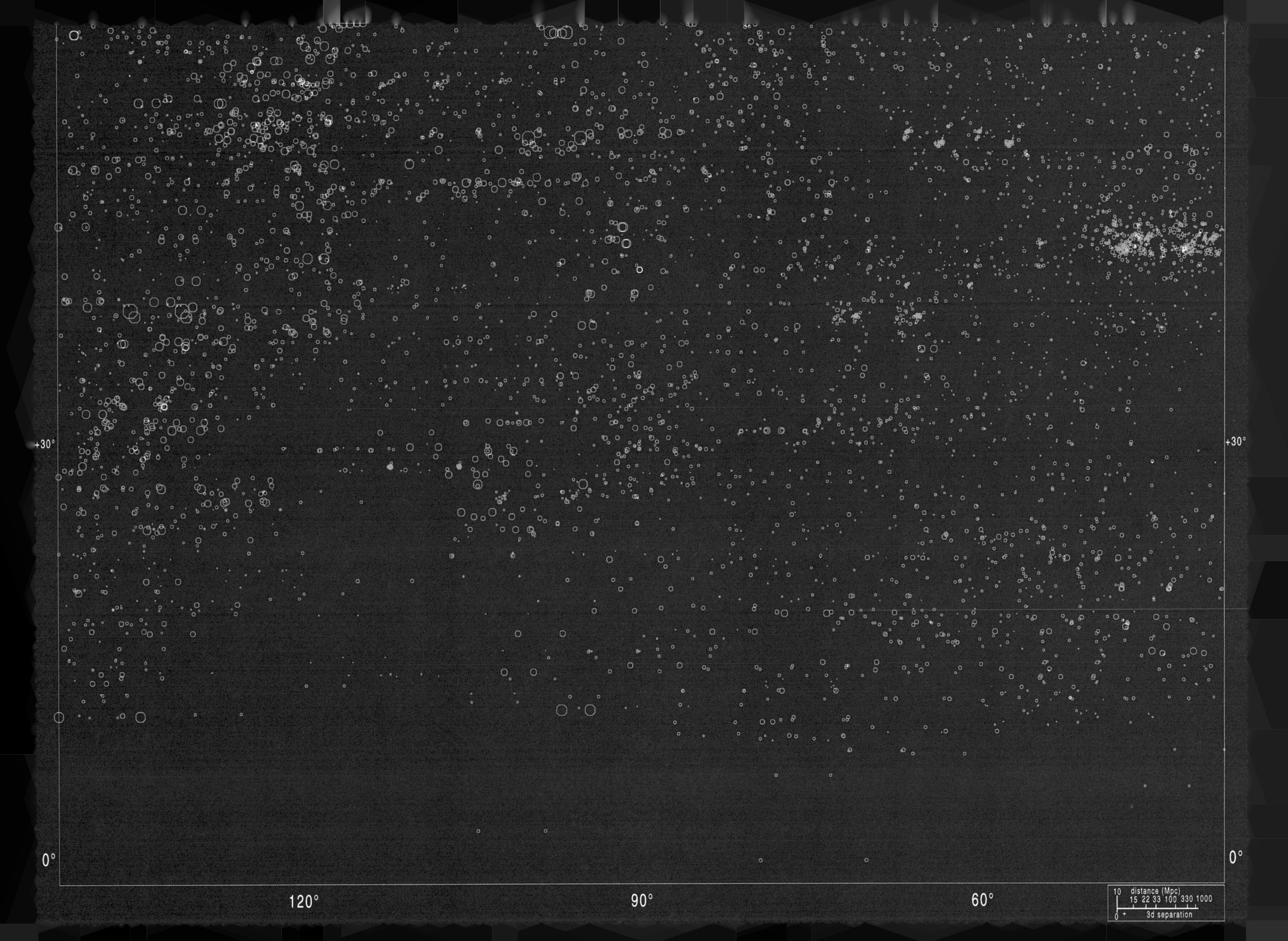
+30°
+30°
0°
0°
120°
90°
60°
distance (Mpc)
10
15 22 33 100 330 1000
0
3d separation

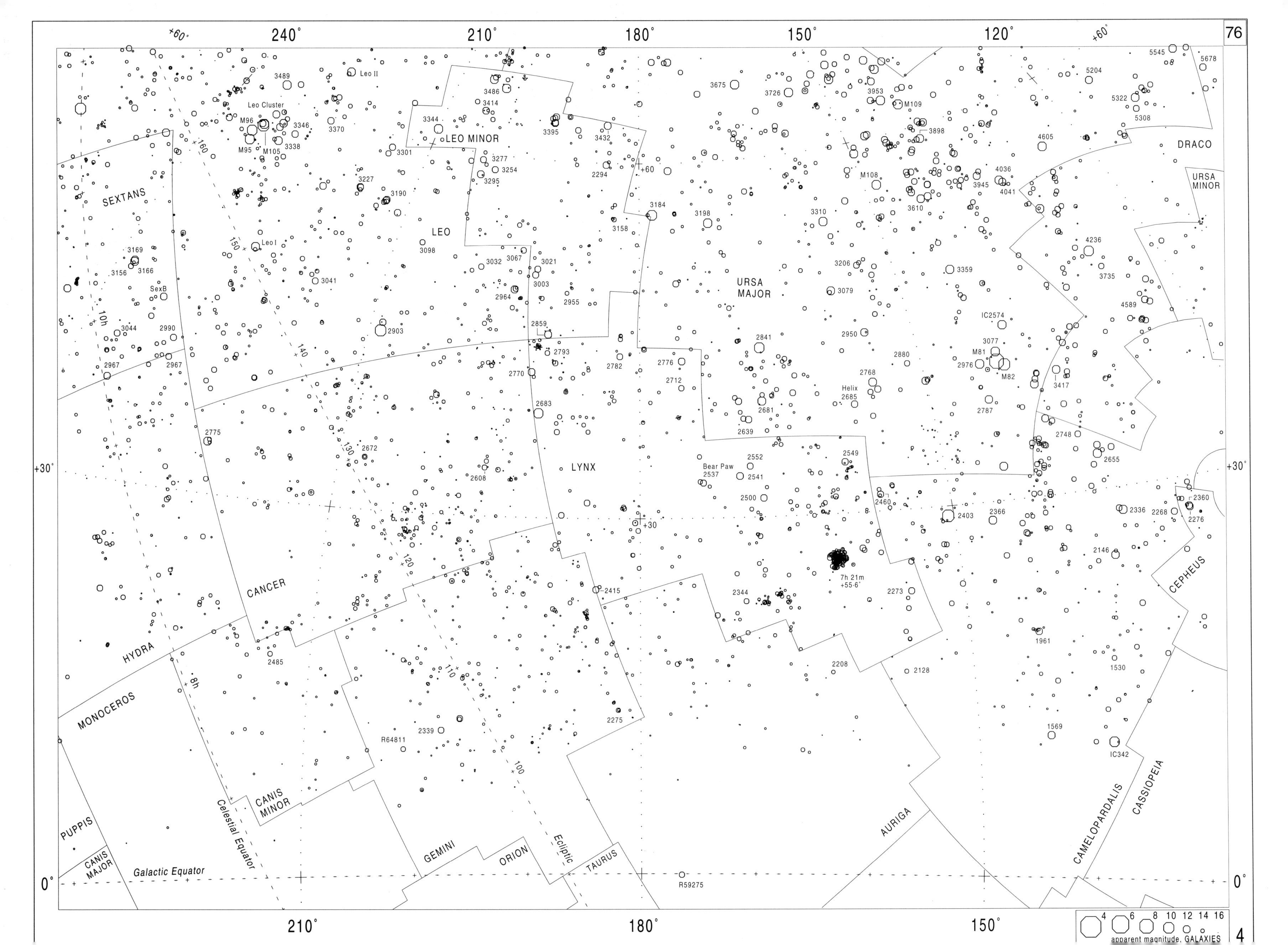

76
DRACO
URSA MINOR
CEPHEUS
CASSIOPEIA
CAMELOPARDALIS
AURIGA
URSA MAJOR
LYNX
LEO MINOR
LEO
SEXTANS
CANCER
HYDRA
MONOCEROS
CANIS MINOR
PUPPIS
CANIS MAJOR
GEMINI
ORION
TAURUS
Ecliptic
Celestial Equator
Galactic Equator
Leo Cluster
Leo I
Leo II
SexB
Helix
Bear Paw
M81
M82
M95
M96
M105
M108
M109
7h 21m +55·6°
240°
210°
180°
150°
120°
+60°
+30°
0°
4 6 8 10 12 14 16
apparent magnitude, GALAXIES
4

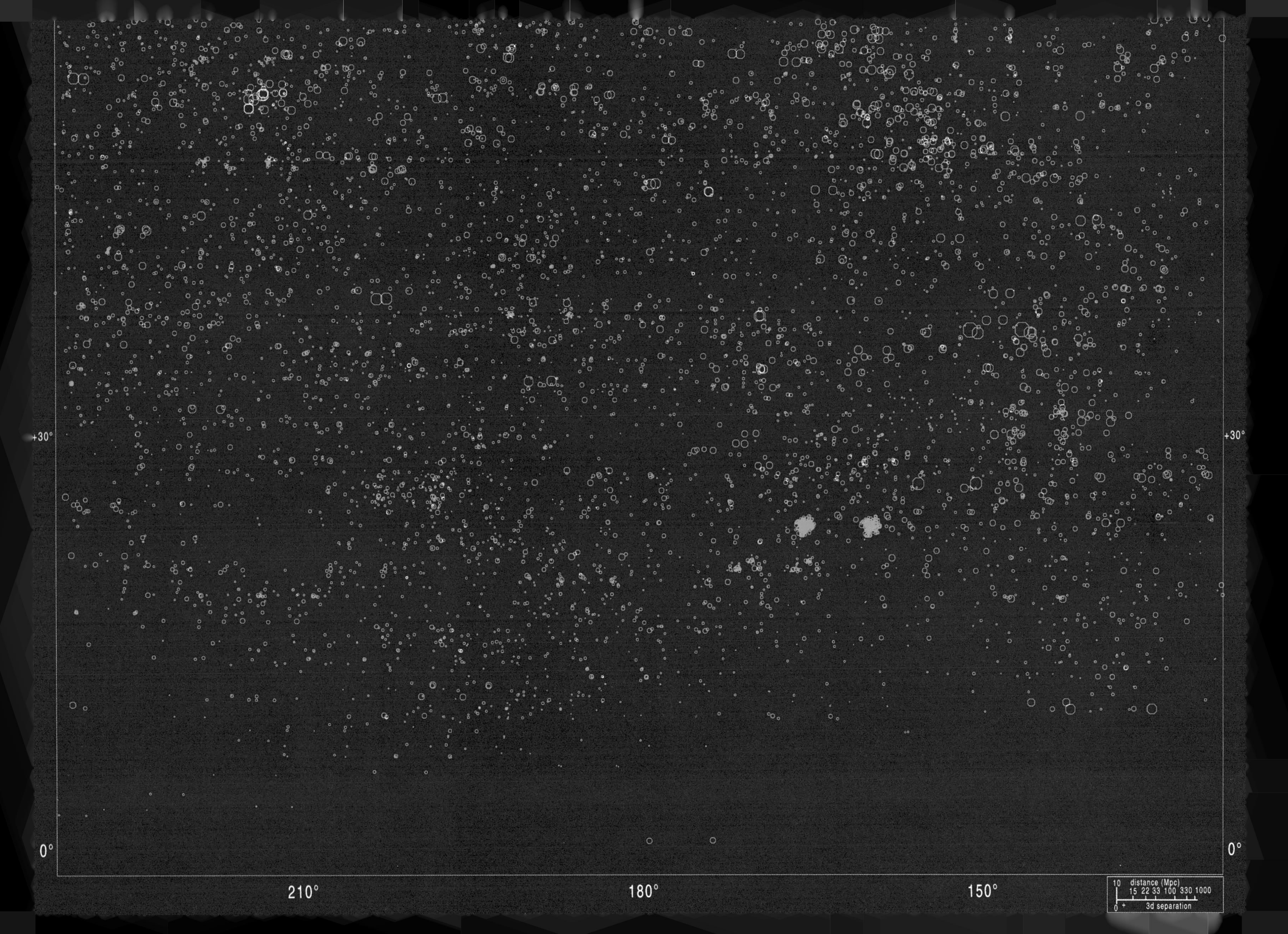

+30°
+30°
0°
0°
210°
180°
150°
10 distance (Mpc)
15 22 33 100 330 1000
0
3d separation

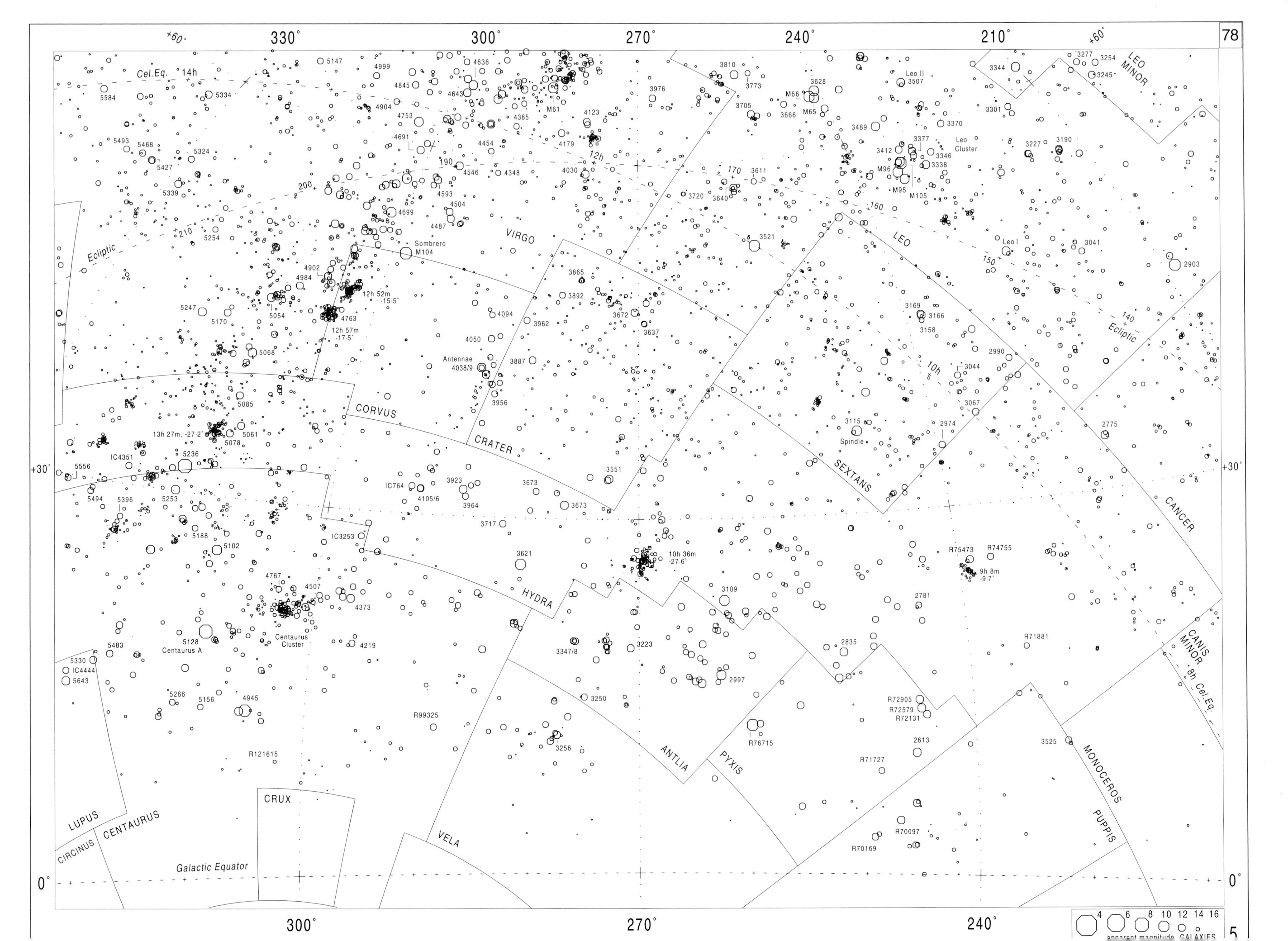
330°
300°
270°
240°
210°
+60°
+30°
0°
LEO MINOR
LEO
Leo I
Leo II
Leo Cluster
M65
M66
M95
M96
M105
CANCER
CANIS MINOR
8h Cel.Eq.
14h Cel.Eq.
12h
10h
Ecliptic
SEXTANS
Spindle
VIRGO
Sombrero M104
M61
CRATER
CORVUS
Antennae 4038/9
HYDRA
12h 52m -15.5°
12h 57m -17.5°
13h 27m, -27.2°
10h 36m -27.6°
9h 8m -9.7°
Centaurus Cluster
Centaurus A
ANTLIA
PYXIS
MONOCEROS
PUPPIS
VELA
CRUX
LUPUS
CENTAURUS
CIRCINUS
Galactic Equator
4 6 8 10 12 14 16
apparent magnitude GALAXIES

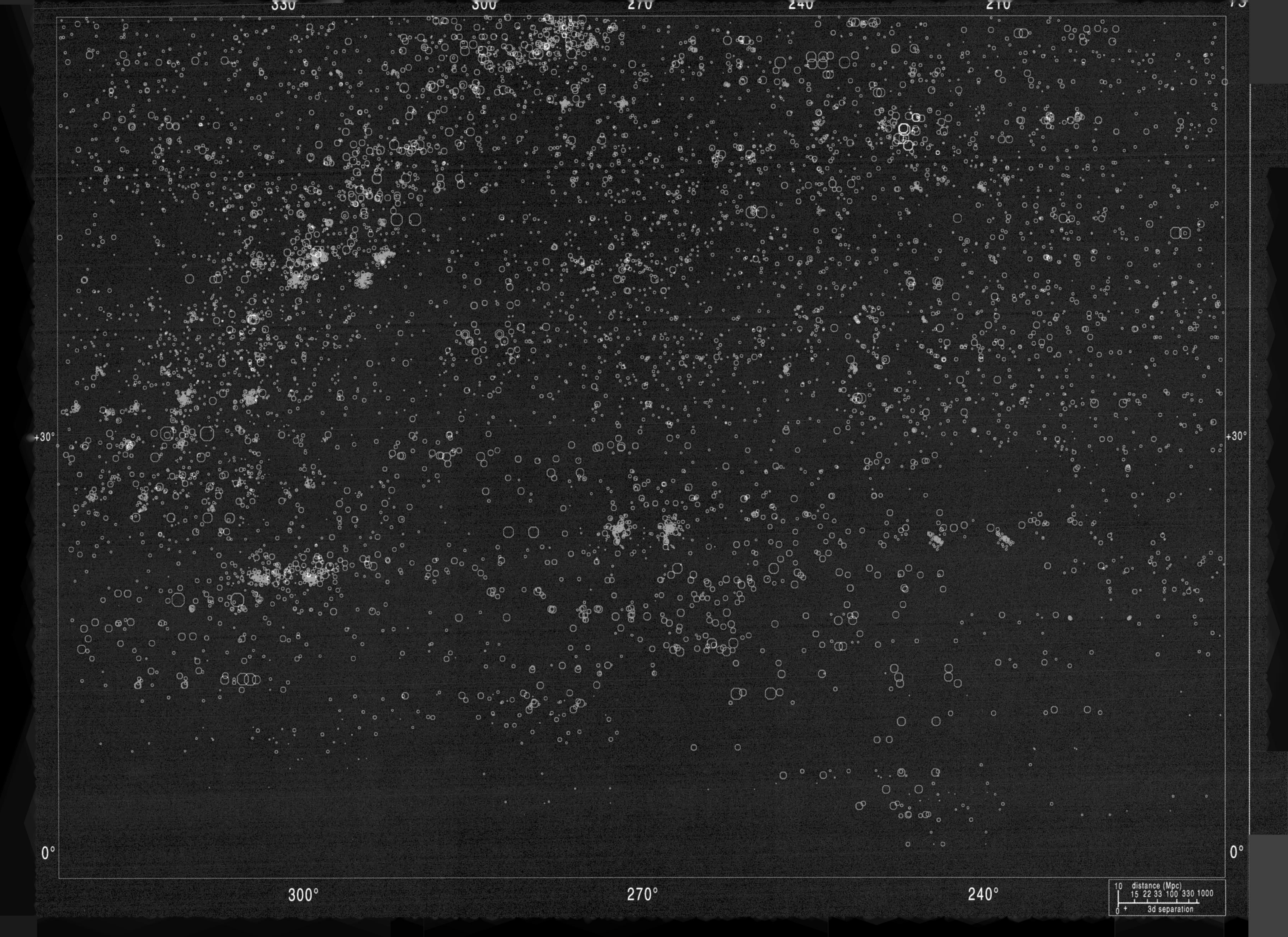

+30°
0°
300°
270°
240°
distance (Mpc)
10 15 22 33 100 330 1000
3d separation

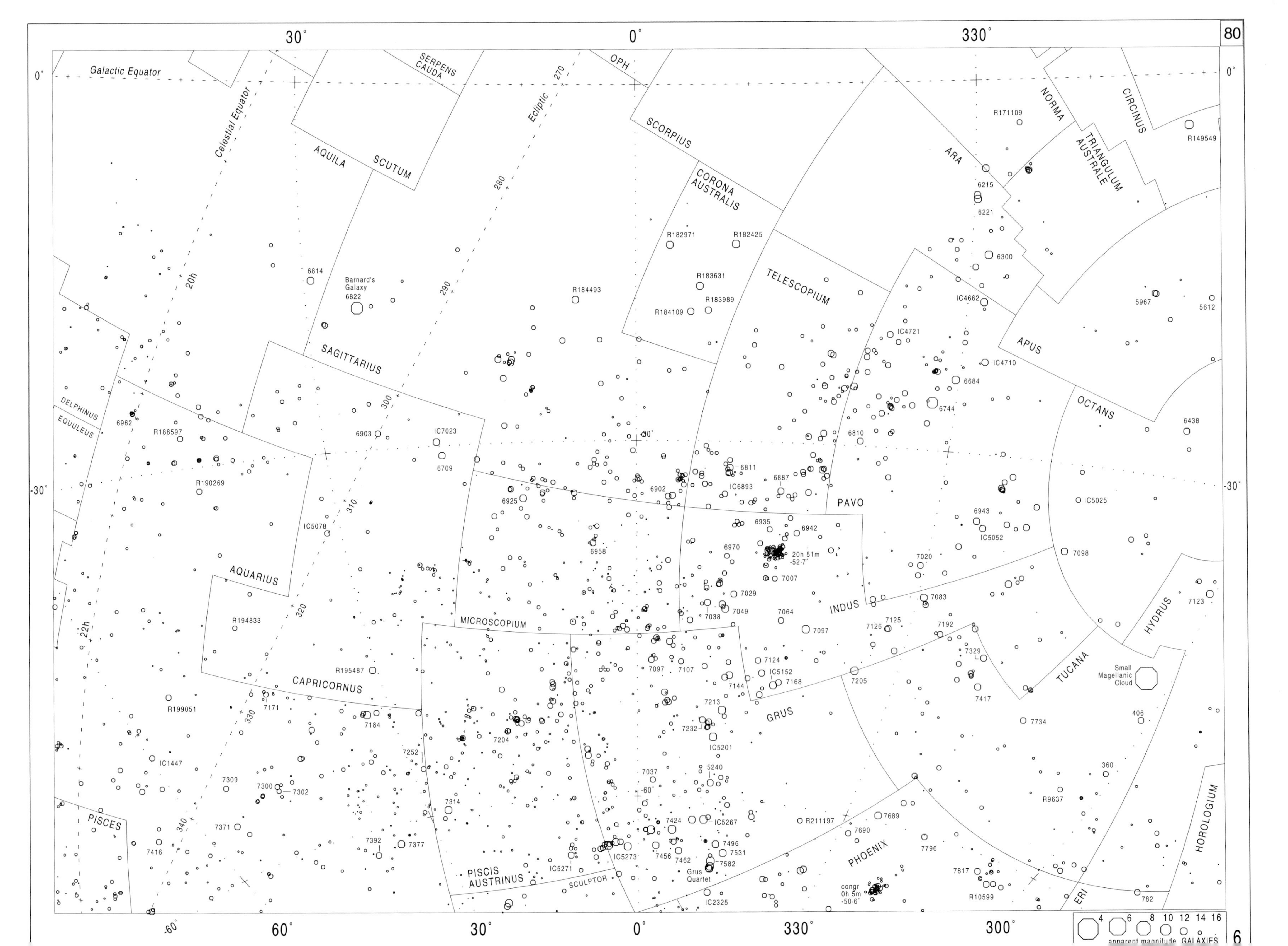

80
Galactic Equator
Celestial Equator
Ecliptic
SERPENS CAUDA
OPH
AQUILA
SCUTUM
SCORPIUS
CORONA AUSTRALIS
ARA
NORMA
TRIANGULUM AUSTRALE
CIRCINUS
TELESCOPIUM
APUS
OCTANS
Barnard's Galaxy
SAGITTARIUS
DELPHINUS
EQUULEUS
PAVO
AQUARIUS
MICROSCOPIUM
INDUS
HYDRUS
TUCANA
Small Magellanic Cloud
CAPRICORNUS
GRUS
PISCES
PISCIS AUSTRINUS
SCULPTOR
Grus Quartet
PHOENIX
HOROLOGIUM
ERI
apparent magnitude GALAXIES
6

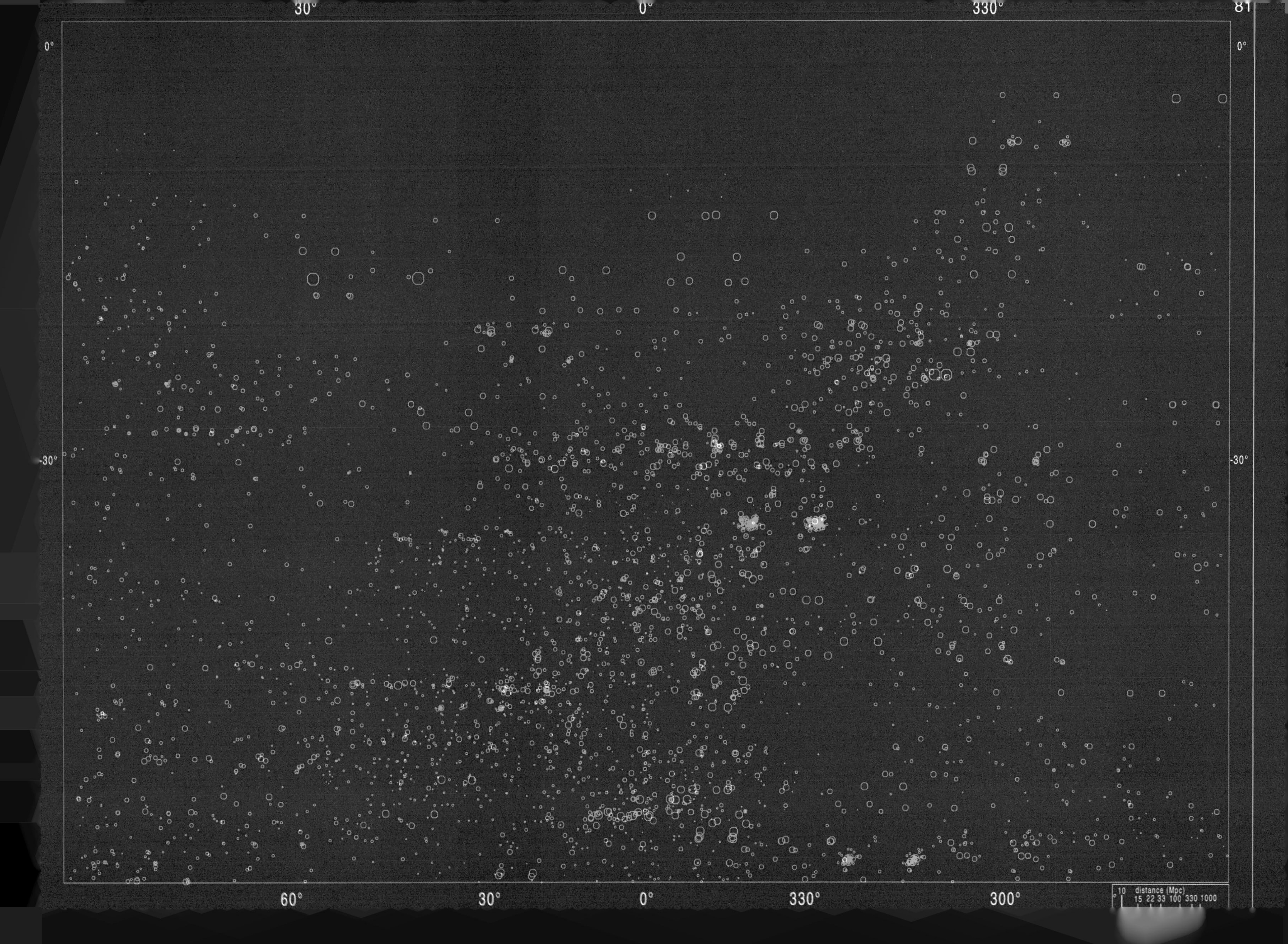
30°
0°
330°
0°
0°
-30°
-30°
60°
30°
0°
330°
300°
distance (Mpc)
10
15 22 33 100 330 1000

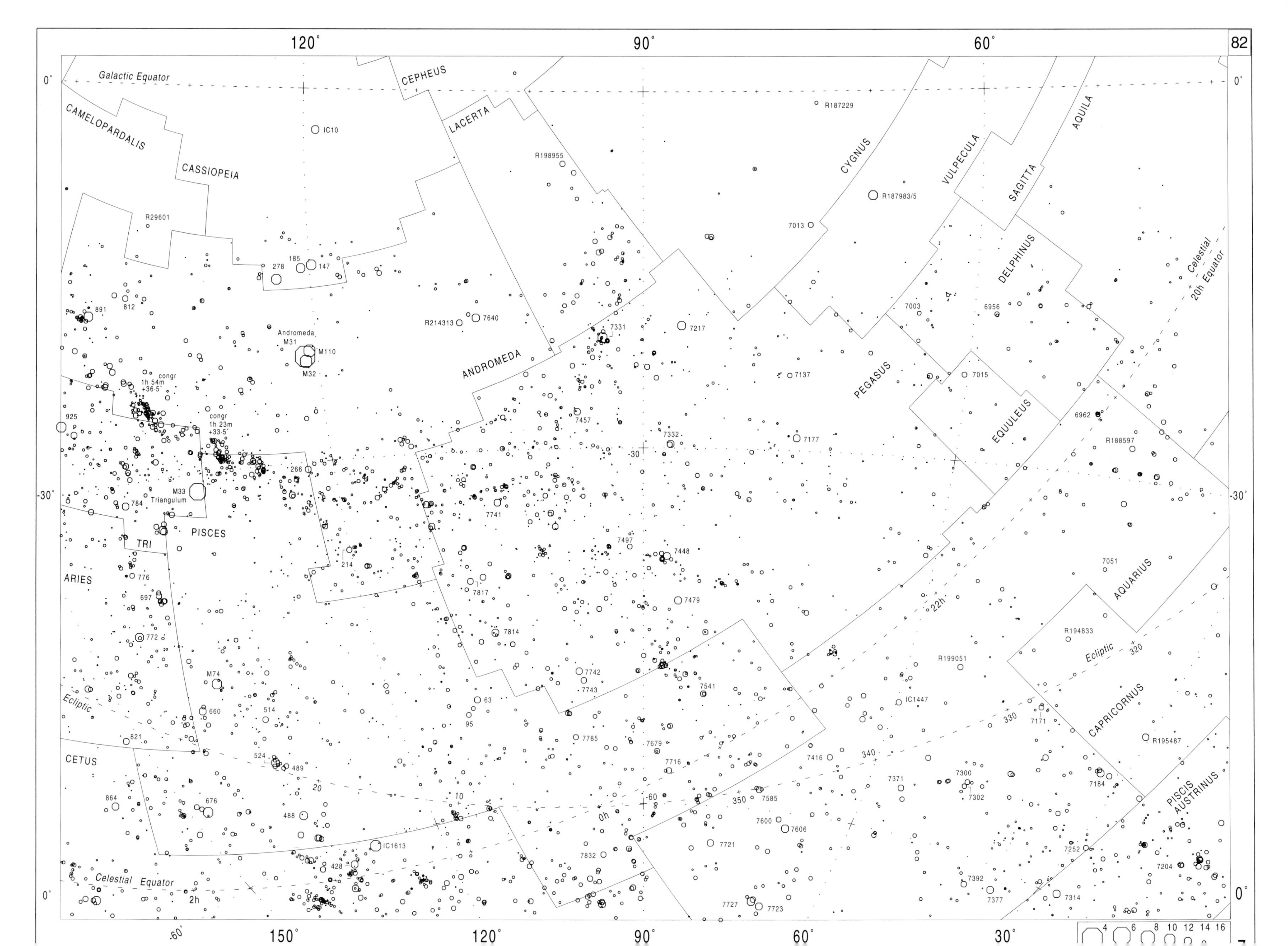

Galactic Equator
CAMELOPARDALIS
CASSIOPEIA
CEPHEUS
LACERTA
CYGNUS
VULPECULA
SAGITTA
AQUILA
DELPHINUS
EQUULEUS
PEGASUS
ANDROMEDA
Andromeda M31
M110
M32
M33 Triangulum
TRI
PISCES
ARIES
CETUS
AQUARIUS
CAPRICORNUS
PISCIS AUSTRINUS
Celestial 20h Equator
Celestial Equator
Ecliptic
congr 1h 54m +36·5°
congr 1h 23m +33·5°
IC10
IC1447
IC1613
R187229
R187983/5
R198955
R29601
R214313
R188597
R194833
R199051
R195487
22h
0h
2h
4 6 8 10 12 14 16

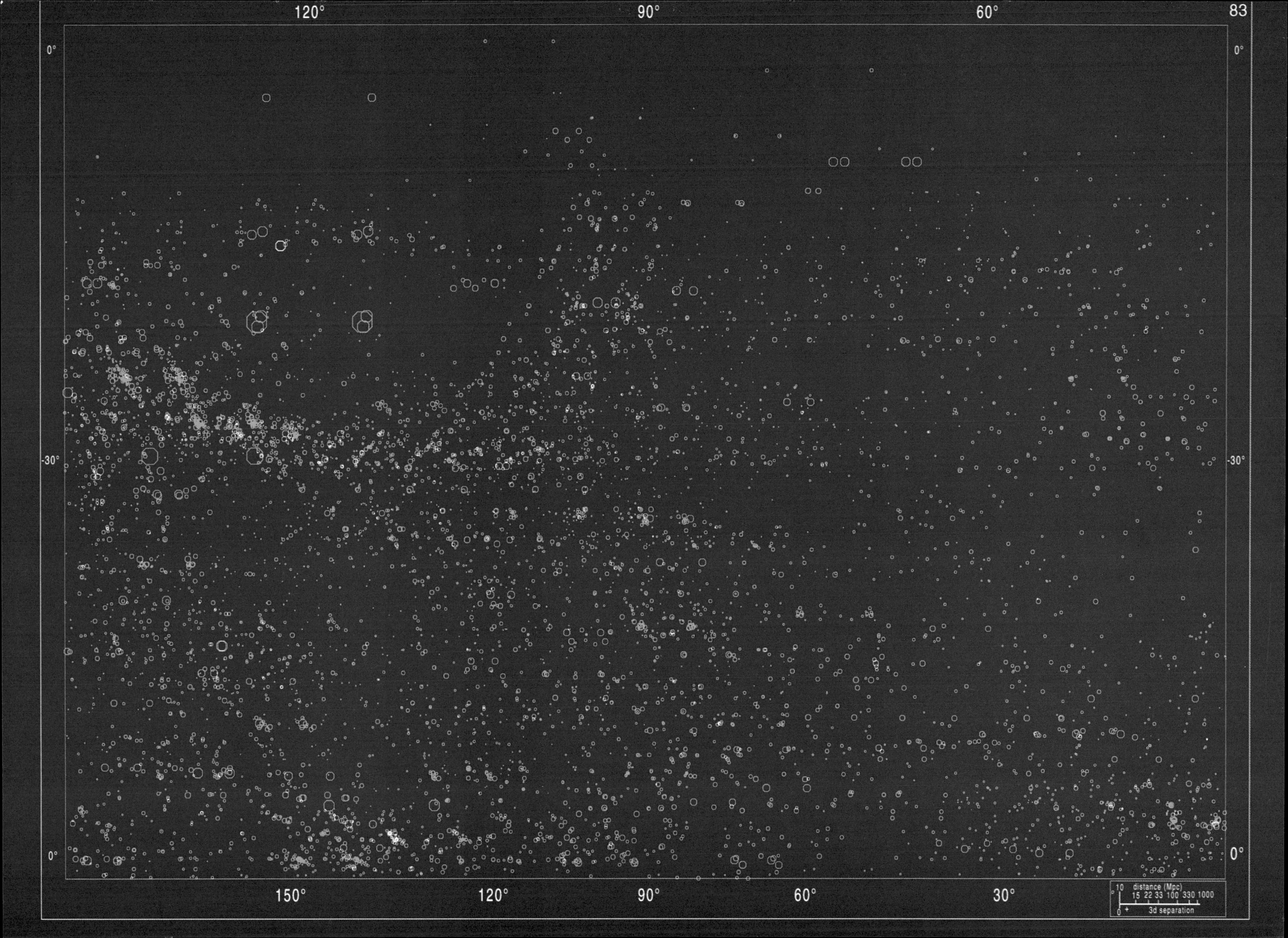

120°
90°
60°
0°
-30°
0°
150°
120°
90°
60°
30°
0°
distance (Mpc)
10
15 22 33 100 330 1000
0
3d separation

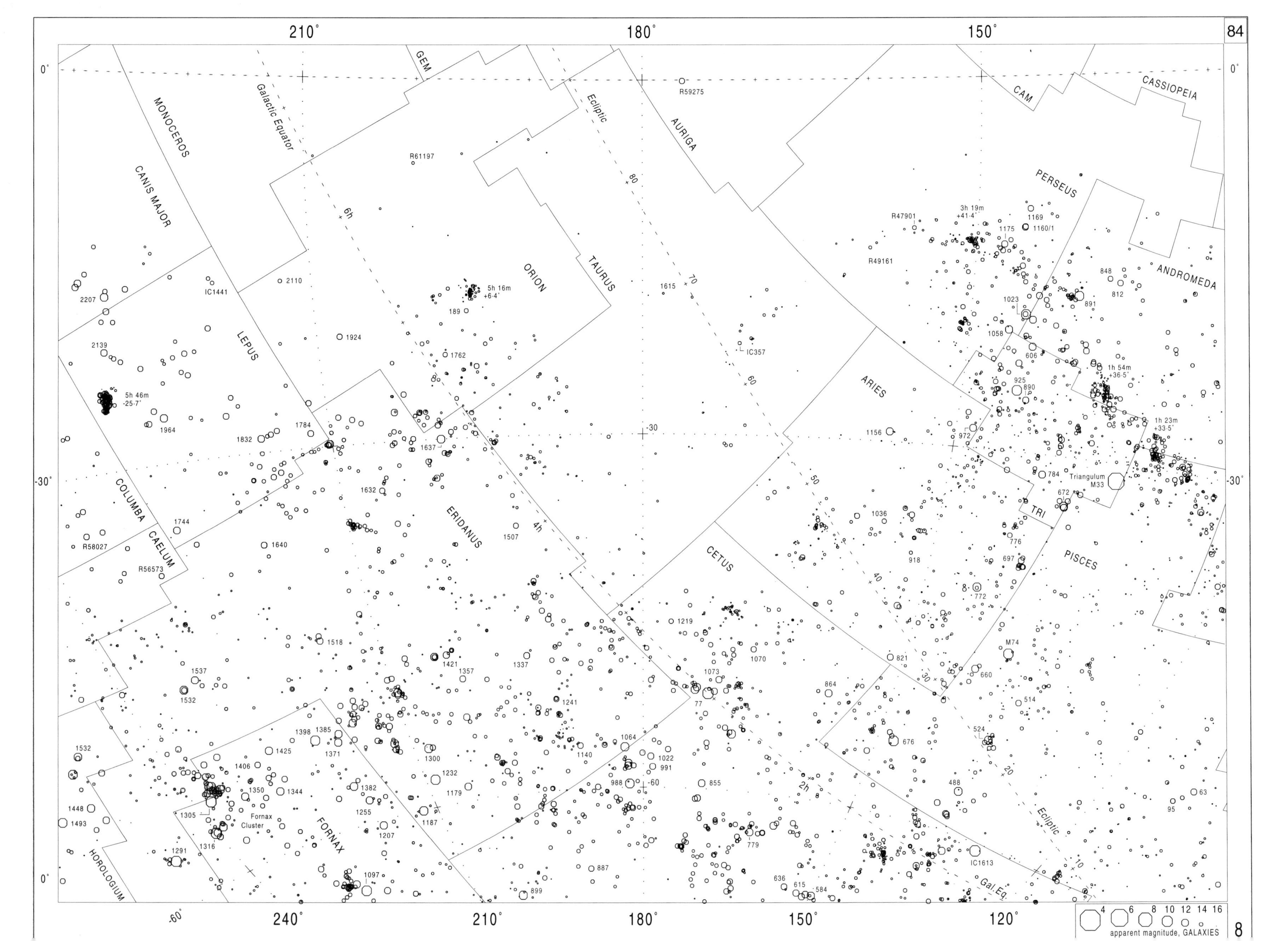

210°
180°
150°
0°
-30°
240°
120°
-60°
MONOCEROS
CANIS MAJOR
LEPUS
COLUMBA
CAELUM
HOROLOGIUM
FORNAX
ERIDANUS
ORION
TAURUS
GEM
AURIGA
CAM
CASSIOPEIA
PERSEUS
ANDROMEDA
TRI
Triangulum
M33
PISCES
ARIES
CETUS
Galactic Equator
Ecliptic
Gal.Eq.
Fornax
Cluster
5h 46m
-25.7°
5h 16m
+6.4°
3h 19m
+41.4°
1h 54m
+36.5°
1h 23m
+33.5°
M74
IC1441
IC357
IC1613
R61197
R59275
R47901
R49161
R58027
R56573
6h
4h
2h
apparent magnitude, GALAXIES

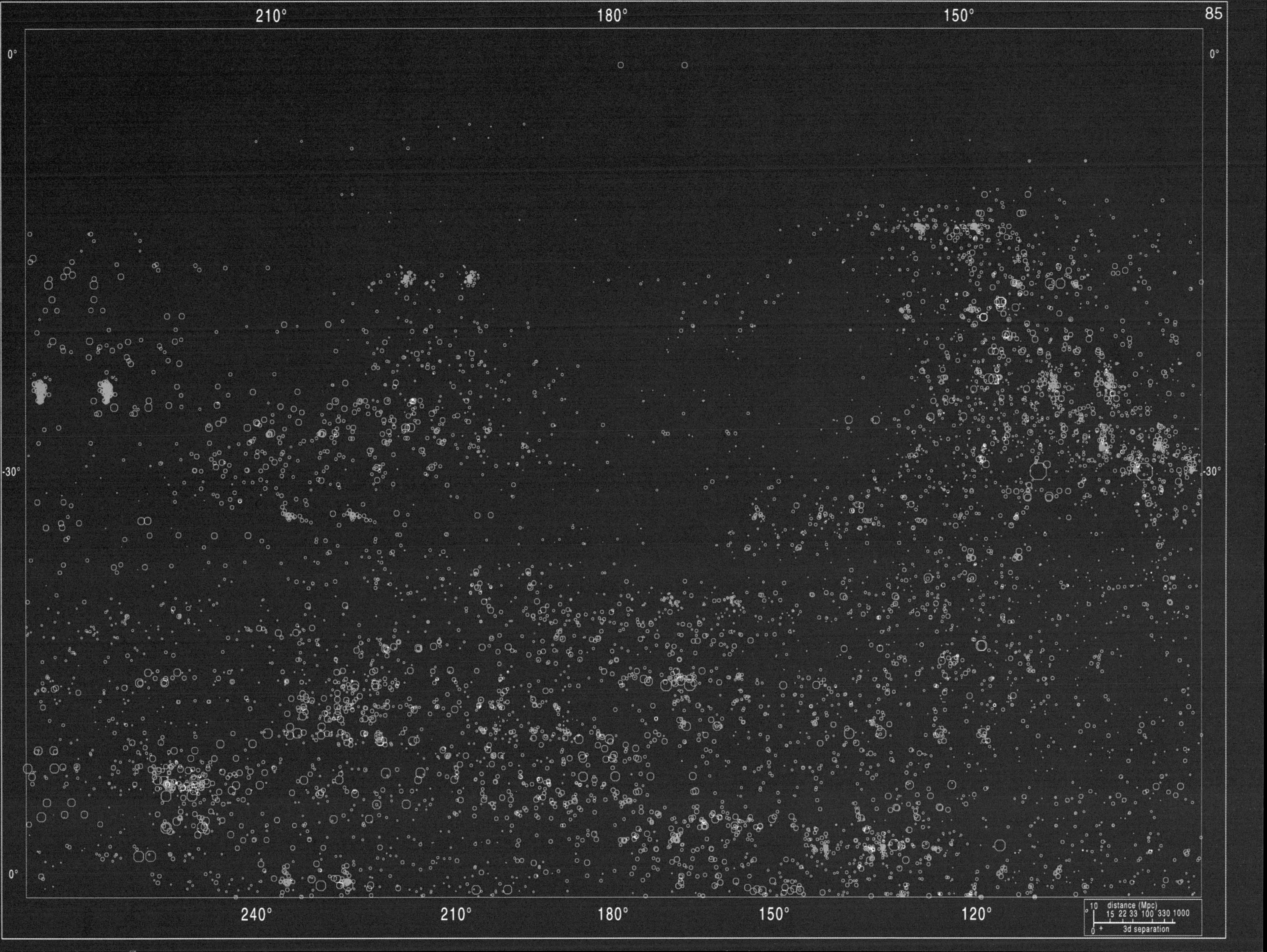
210°
180°
150°
0°
-30°
0°
240°
210°
180°
150°
120°
0°
-30°
0°
10 distance (Mpc)
15 22 33 100 330 1000
0
3d separation

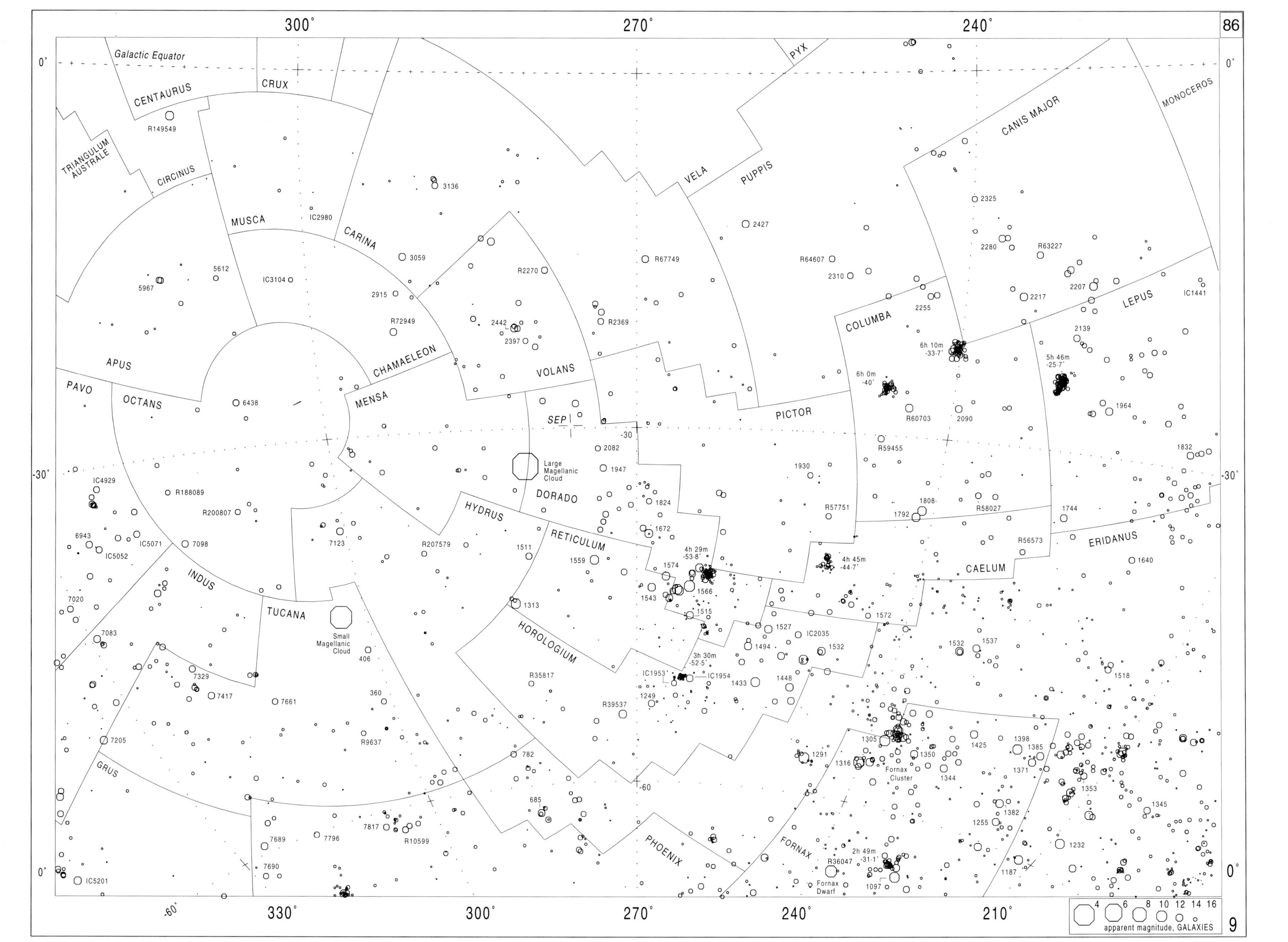

300°
270°
240°
0°
Galactic Equator
CENTAURUS
CRUX
PYX
MONOCEROS
CANIS MAJOR
TRIANGULUM AUSTRALE
CIRCINUS
MUSCA
CARINA
VELA
PUPPIS
LEPUS
COLUMBA
APUS
PAVO
OCTANS
MENSA
CHAMAELEON
VOLANS
SEP
PICTOR
Large Magellanic Cloud
DORADO
HYDRUS
RETICULUM
CAELUM
ERIDANUS
INDUS
TUCANA
Small Magellanic Cloud
HOROLOGIUM
GRUS
PHOENIX
FORNAX
Fornax Cluster
Fornax Dwarf
R149549
3136
2427
2325
2280
R63227
5612
IC2980
5967
IC3104
3059
R2270
R67749
R64607
2310
2207
2217
IC1441
2915
R72949
2442
2397
R2369
2255
6h 10m -33·7˚
5h 46m -25·7˚
2139
6h 0m -40˚
1964
6438
R60703
2090
-30
2082
R59455
1832
1947
1930
-30°
IC4929
R188089
R200807
1824
1808
R57751
R58027
1792
1744
6943
IC5071
IC5052
7098
7123
R207579
1511
1672
1559
4h 29m -53·8˚
1574
4h 45m -44·7˚
R56573
1640
7020
1313
1543
1566
1515
1572
7083
406
1527
IC2035
1532
1537
1494
7329
7417
7661
360
3h 30m -52·5˚
IC1953
IC1954
1433
1448
1518
1249
R35817
R39537
7205
R9637
1305
1291
1316
1350
1425
1398
1385
1371
1344
1353
782
-60
685
1382
1255
1345
7817
R10599
7689
7796
2h 49m -31·1˚
R36047
1232
7690
1187
IC5201
1097
-60°
330°
210°
4 6 8 10 12 14 16
apparent magnitude, GALAXIES

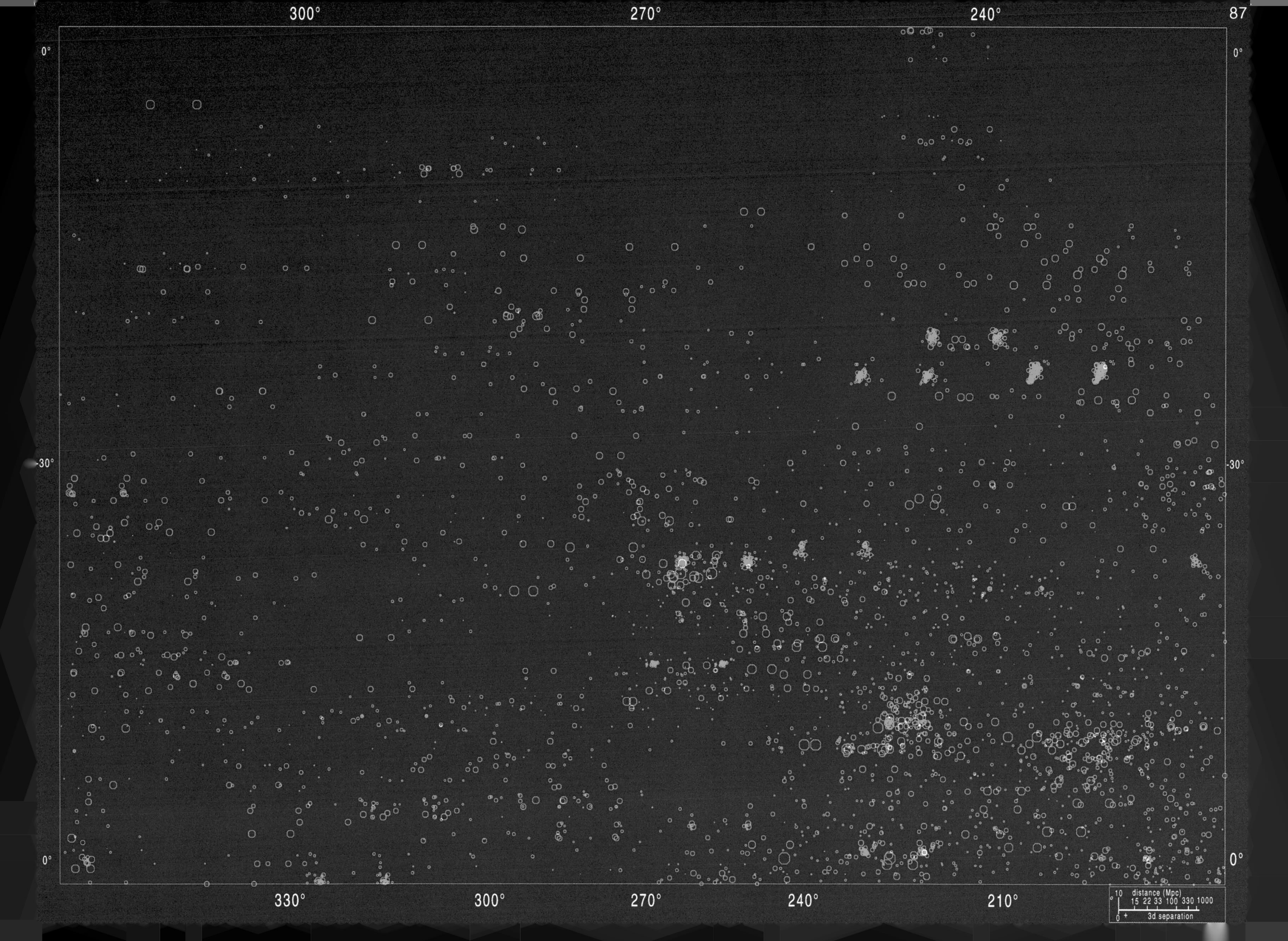
300°
270°
240°
0°
-30°
0°
330°
300°
270°
240°
210°
distance (Mpc)
10 15 22 33 100 330 1000
0 +
3d separation

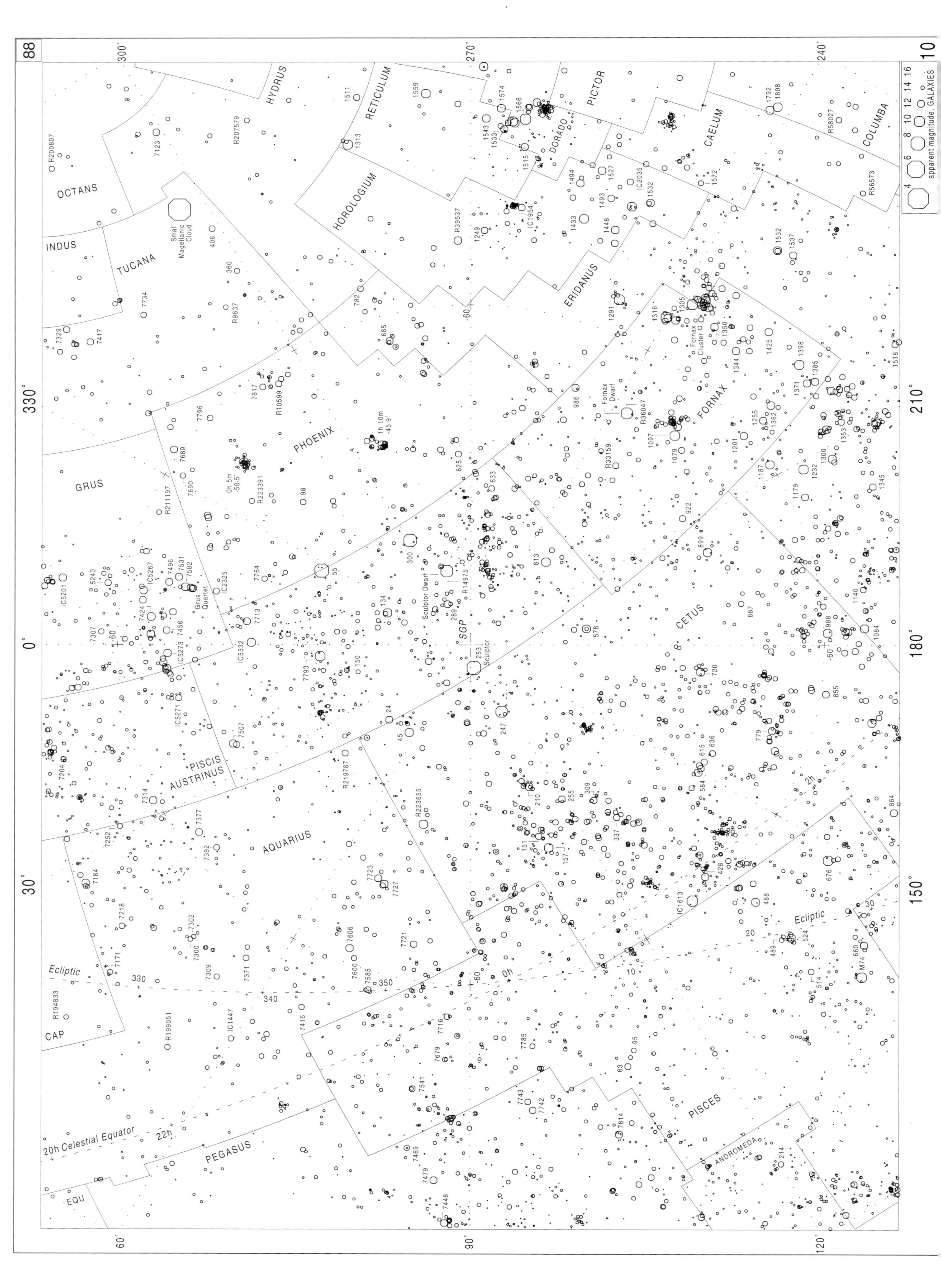

88
10
OCTANS
INDUS
TUCANA
Small Magellanic Cloud
HYDRUS
RETICULUM
HOROLOGIUM
DORADO
PICTOR
CAELUM
COLUMBA
ERIDANUS
GRUS
PHOENIX
Grus Quartet
Sculptor Dwarf
SGP
Sculptor
FORNAX
Fornax Cluster
Fornax Dwarf
CETUS
PISCIS AUSTRINUS
AQUARIUS
CAP
Ecliptic
Celestial Equator
PEGASUS
PISCES
ANDROMEDA
EQU
M74
apparent magnitude, GALAXIES
4 6 8 10 12 14 16
300°
270°
240°
330°
0°
30°
210°
180°
150°
60°
90°
120°
0h 5m -50.5°
1h 10m -45.9°

30°
0°
330°
89
60°
300°
90°
270°
120°
240°

# Special Views

The eight 'special view' sterographs shown below concentrate on three subjects, prper motion in stars, comprehensive views, and galaxy cluster detail. They have been prepared from a variety of sources.

Views 1-3 show proper motion in stars, and have been plotted from the *Yale Bright Star Catalogue* (1991), with distances calculated from spectroscopic parallax. The circular object symbol shows the position of the star (epoch 2000). The line running into the object plot is extrapolated back from the current proper motion data, and shows the notional track of the star's proper motion over the last 100 000 years had it been moving through space in a straight lne on its current course and at its current speed over that period.

View 1 shows the proper motion of objects in the Hyades (at about 50 pc) abd the Pleiades (at about 100 pc) star clusters. A method of distance determination nort otherwise discussed in the present work is the 'moving cluster' technique, that assumes the objects in a star cluster such as the Hyades are moving on parallel tracks, and calculates the distance from the perspective, the degree of apparent divergence on convergence seen in those parallel tracks. That the orange giant star Aldebaran is not a member of the Hyades cluster is apparent from its nearer distance and quite different track.

The 'proper motion' of a star is a measurement of apparent motion from two different sources, the motion of of the star and the motion of the Sun. That the Sun has its own individual proper motion in one direction (towards Ophiuchus) is seen from the way other stars show an apparent motion in the opposite overall direction. This apparent proper motion is most visible at about ninety degrees to the direction of the Sun's motion, in a 'left hand window' towards Perseus (view 2) and in a 'right hand window' towards Crux (view 3).

Views 4 and 5 combine stars, galactic non-stellar objects, and external galaxies, with representation of morphological type. Stars to a limit at apparent magnitude 4.49 are shown with symbol size representing nearness, and are plotted from the *Yale Bright Star Catalogue* (1991). External galaxies are plotted from the *CfA Redshift Catalogue* (1998). Galactic and Magellanic non-stellar objects are plotted from a hybrid catalogue prepared by the authors from a variety of sources. Of all the sterographs in the present work, views 4 and 5 cover the greatest distance range and the widest range of objects. Te star-plots represent objects at a close range, external galaxies are objects at a far distant range, and the galactic non-stellar objects are seen in the near to intermediate range, for the most part distributed along the nearby arms in the spiral structure of the Milky Way.

Views 6-8 show groupings of external galaxies at a slightly enlarged scle to those seen in the main series of galactic maps: view 6 shows the cluster in Coma Berenices (galaxy map 1), view 7a detail from the filament structure across Andromeda (maps 7 and 8), view 8 shows the cluster in Centaurus (map 5.) The special views differ from the main series of galaxy maps by using a greater distance to the plane of the page, 50 Mpc in place of 10 Mpc. This opens up the particular fields shown in the sterographs, and enables them to be seen in grater separation and detail.

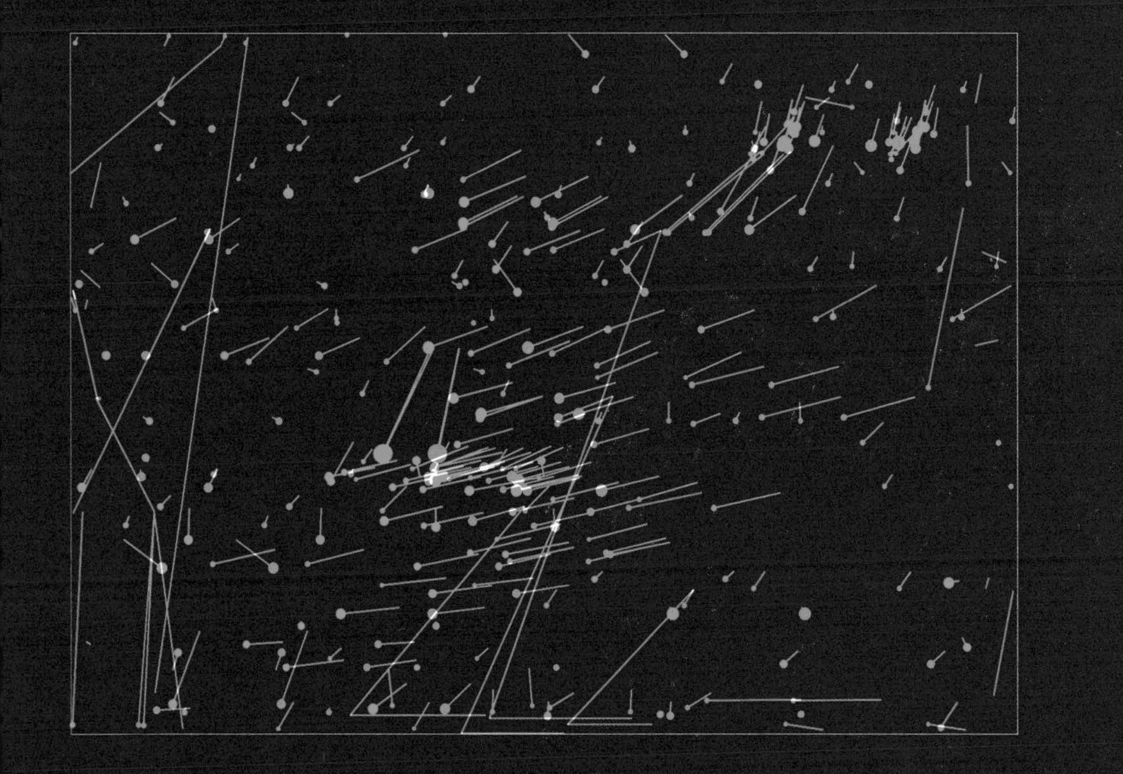

## Special views 1-3

## Stellar proper motion over 100,000 years

All fields are 40° wide.

Top left : Some star clusters are clearly moving as a whole. This view shows the motion of the Hyades and Pleiades.

The relative motion of the sun (130,000km/hr in the direction of Ophiuchus) causes an apparent motion or 'streaming effect' of the surrounding stars.

Top right: The streaming effect as viewed from LH solar window. Perseus appears as moving cluster lower left.
Direction: galactic longitude 140°.

Bottom right: Streaming effect, viewed from RH solar window. Crux is near the centre.
Direction: galactic longitude 300°.

The galactic plane runs left right at centre of the two images on the right.

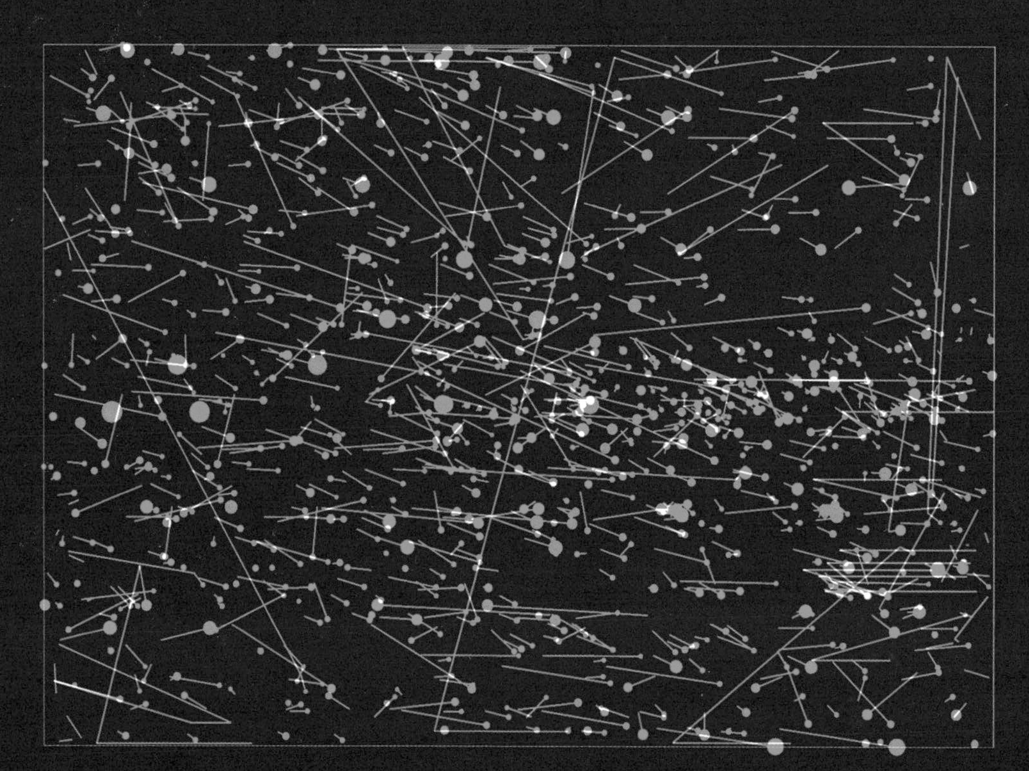

Special View 4.
Wide angle (320deg) view of the Milky Way Galaxy. Objects near to far: stars (filled circles), clusters (dotted circles), gas clouds (squares), and galaxies showing different morphological types from elliptic through to full spiral.
For the stars, size shows nearness, otherwise size shows brightness.
This view clearly shows how the dark gas in the Milky Way obscures the light from the more distant galaxies. The Magellanic Clouds are seen right of centre below the galactic plane.

Special View 5.
Wide angle (320deg) view of the Milky Way Galaxy, looking away from the galactic centre.  See special view 4 for notes on objects.

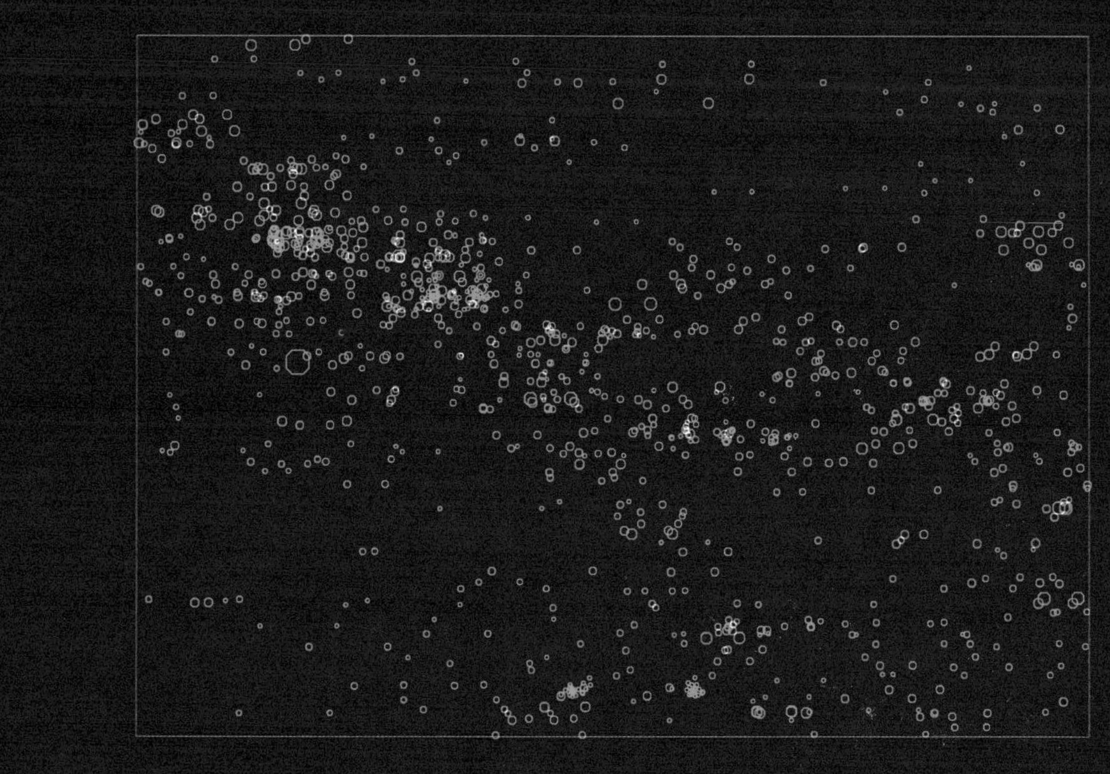

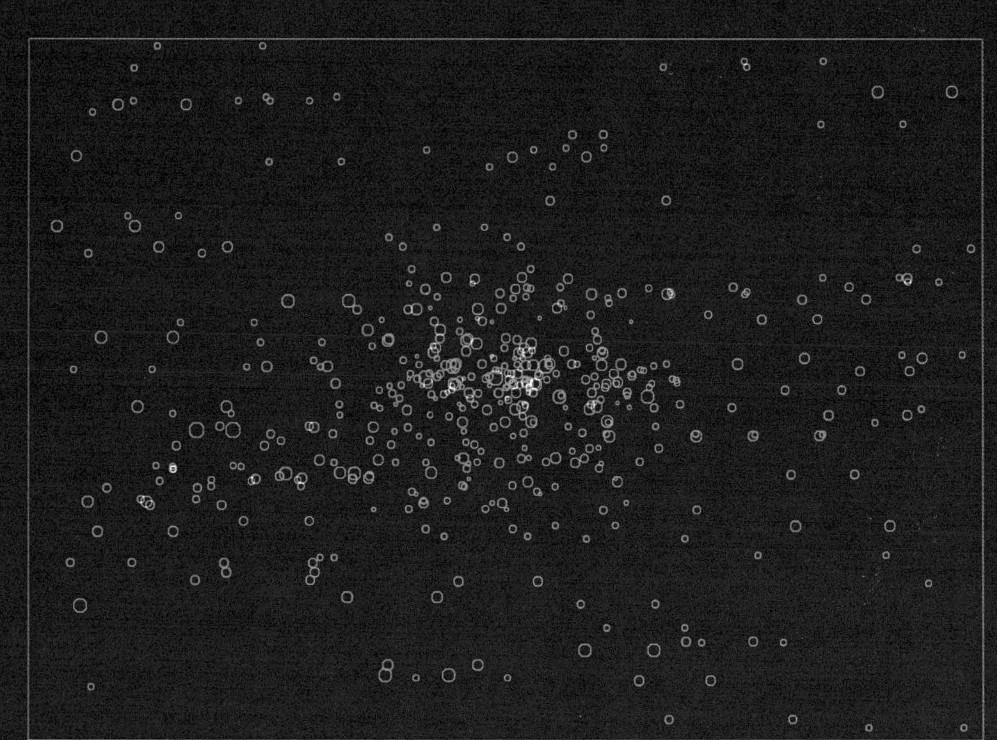

**Special views 6-8**

**Supergalactic structure and clustering**

Above left: Cluster in Coma Berenices.
- Glo 58° Gla 88°. Field width 3.5°. Plane of paper is 50MPc.

Above right: 'Twisted Rope' of filament of galaxies near Andromeda. - Glo 123° Gla -32°. Field width 23°. Plane of paper is 50Mpc.

Below right: Cluster in Centaurus.
- Glo 302° Gla 21°. Field width is 6°. Plane of paper is 20Mpc.

## Sources

*The Bright Star Catalogue*, ed. Dorrit Hoffleit, Carlos Jaschek and Wayne Warren Jr, 5th edition (machine readable), Department of Astronomy, Yale University, 1991

*CfA Redshift Catalogue of Galaxies*, ed. John Huchra, M. Geiler *et al.* (machine readable), Harvard-Smithsonian Center for Astrophysics, Cambridge Mass., 1998

*The Hipparcos and Tycho Catalogues* (machine readable), European Space Agency, ESTEC, Noordwijk, 1997

*NGC 2000.0, The Complete New General Catalogue and Index Catalogues of Nebulae and Star Clusters by J.L.E. Dreyer*, ed. Roger W. Sinnott (machine readable), Sky Publishing Corporation, Cambridge Mass., 1989

*Preliminary Version of the Third Catalogue of Nearby Stars*, ed. W. Gliese and H. Jahreiss (machine readable), Astron. Rechen-Institut, Heidelburg, 1991

*Sky Catalogue 2000.0, Volume 1: Stars to magnitude 8.0*, ed. Alan Hirshfeld and Roger W. Sinnot, Cambridge University Press/Sky Publishing Corporation, 1981

*Sky Catalogue 2000.0, Volume 2: Double Stars, Variable Stars and Nonstellar Objects*, ed. Alan Hirshfeld and Roger W. Sinnot, Cambridge University Press/Sky Publishing Corporation, 1985

*A Supplement to the Bright Star Catalogue*, ed. Dorrit Hoffleit, Michael Saladyga and Peter Wlasuk (machine readable), Yale University Observatory, 1983

Eugène Delaporte, *Délimitation Scientifique des Constellations (Tables et Cartes)* [The Report of Commission 3], International Research Council, International Astronomical Union, Cambridge University Press, 1930

James B. Kaler, *Stars and their Spectra*, Cambridge University Press, 1989

Paul Kunitzsch and Tim Smart, *Short Guide to Modern Star Names and their Derivations,* Otto Harrassowitz, Wiesbaden, 1986

Hugh C. Maddocks, *Deep-Sky Name Index 2000.0*, Foxon Maddocks Associates, Reston, Va., 1991

A. P. Norton, *Norton's Star Atlas*, 18th edition, ed. Ian Ridpath, Longman, Harlow, 1989

Wil Tirion, *Sky Atlas 2000.0*, Cambridge University Press/Sky Publishing Corporation, 1981

R. Brent Tully and J. Richard Fisher, *Nearby Galaxies Atlas*, Cambridge University Press, 1987

## Acknowledgements

The authors would like to thank Stephen Adamson, Eunice Cox, Dr F. van Leeuwen, Dr Robert Massey, Dr Paul Murdin, Caroline Rayner, and the Library at the Institute of Astronomy, University of Cambridge, for their help and assistance in preparing this Atlas. Responsibility for errors remains with the authors.